Sheila D. Petcavage
Instructor

Business

Western Campus

11000 Pleasant Valley Road

Parma, Ohio 44130-5199

216•987•5571 FAX 216•987•5013
email: sheila.petcavage@tri-c.cc.oh.us

Cuyahoga Community College

D1306409

PURCHASING FOR BOTTOM LINE IMPACT

IMPROVING THE ORGANIZATION THROUGH STRATEGIC PROCUREMENT

THE NAPM PROFESSIONAL DEVELOPMENT SERIES

Michiel R. Leenders
Series Editor

Volume I

VALUE-DRIVEN PURCHASING
Managing the Key Steps in the Acquisition Process
Michiel R. Leenders
Anna E. Flynn

Volume II

MANAGING PURCHASING
Making the Supply Team Work
Kenneth H. Killen
John W. Kamauff

Volume III

VALUE-FOCUSED SUPPLY MANAGEMENT
Getting the Most Out of the Supply Function
Alan R. Raedels

Volume IV

PURCHASING FOR BOTTOM LINE IMPACT
Improving the Organization Through Strategic Procurement
Lisa M. Ellram
Laura M. Birou

PURCHASING FOR BOTTOM LINE IMPACT

IMPROVING THE ORGANIZATION THROUGH STRATEGIC PROCUREMENT

Volume IV
The NAPM Professional Development Series

Lisa M. Ellram
Arizona State University
and
Laura M. Birou
The George Washington University

Tempe, Arizona

McGraw-Hill

New York San Francisco Washington, D.C. Auckland Bogotá
Caracas Lisbon London Madrid Mexico City Milan
Montreal New Delhi San Juan Singapore
Sydney Tokyo Toronto

McGraw-Hill

A Division of The McGraw-Hill Companies

Library of Congress Cataloging-in-Publication Data

Ellram, Lisa M.
    Purchasing for bottom line impact: improving the organization
through strategic procurement/Lisa M. Ellram, Laura M. Birou.
        p. cm.—(The NAPM professional development series; v. 4)
    Includes bibliographical references and index.
    ISBN 0-7863-0217-8
    1. Industrial Procurement. I. Birou, Laura M. II. Title.
III. Series.
HD39.5.E39   1995
658.7'2—dc20   94–49389

Printed in the United States of America

3 4 5 6 7 8 9 BKP/BKP 9 0 2 1 0 9 8

We dedicate this book to our parents,
Aime and Ergav N. Ellram, and
Beverly and Ronald Birou;
for all you have given us, taught us,
and been to us. Thank you.

SERIES OVERVIEW

The fundamental premise for this series of four textbooks is that effective purchasing or supply management can contribute significantly to organizational goals and strategies. This implies that suppliers and the way organizations relate to them are a major determinant of organizational success.

Differences do exist between public and private procurement, between purchasing for service organizations, manufacturers, retailers, distributors, and resource processors; between supplying projects, research and development, job shops, and small and large organizations across a host of industries, applications, and needs. Nevertheless, research has shown much commonality in the acquisition process and its management.

These four textbooks, therefore, cover the common ground of the purchasing field. They parallel the National Association of Purchasing Management (NAPM) Certification Program leading to the C.P.M. designation. They also provide a sound, up-to-date perspective on the purchasing field for those who may not be interested in the C.P.M. designation.

The textbooks are organized into the following four topics:

1. *Value-Driven Purchasing: Managing the Key Steps in the Acquisition Process*
2. *Managing Purchasing: Making the Supply Team Work*
3. *Value-Focused Supply Management: Getting the Most Out of the Supply Function*
4. *Purchasing for Bottom Line Impact: Improving the Organization Through Strategic Procurement*

Volume I, *Value-Driven Purchasing: Managing the Key Steps In the Acquisition Process,* focuses on the standard acquisition process and its major steps, ranging from need recognition and purchase requests to supplier solicitation and analysis, negotiation, and contract execution, implementation, and administration.

Volume II, *Managing Purchasing: Making the Supply Team Work,* focuses on the administrative aspects of the purchasing department, including the development of goals and objectives, maintenance of files and records, budgeting, and evaluating performance. It also discusses the personnel issues of the function: organization, supervision and delegation of work, evaluating staff performance, training staff, and performance difficulties.

Volume III, *Value-Focused Supply Management: Getting the Most Out of the Supply Function,* commences with identifying material flow activities and decisions, including transportation, packaging requirements, receiving, and interior materials handling. It goes on to cover inventory management and concludes with supply activities such as standardization, cost reduction, and material requirements planning.

Volume IV, *Purchasing for Bottom Line Impact: Improving the Organization Through Strategic Procurement,* begins with purchasing strategies and forecasting. This is followed by internal and external relationships, computerization, and environmental issues.

It is a unique pleasure to edit a series of textbooks like these with a fine group of authors who are thoroughly familiar with the theory and practice of supply management.

Michiel R. Leenders
Series Editor

PREFACE

This book, like the other three in this series, was based on the sixth edition of the CPM Study Guide. This 1992 NAPM publication was intended to assist those preparing for the Certified Purchasing Manager (CPM) examinations. The Guide and its predecessors are the collective work of a large number of purchasing academic and professionals who acted as editors, authors, and reviewers. For the sixth edition these tasks fell to Eugene Muller and Donald W. Dobler, editors; and Harry Robert Page and Eberhard E. Scheuing, consulting editors. We would like to thank the many authors as well as the editorial review board for their fine work.

We would also like to thank Michiel Leenders, the series editor for his timely input, review, advice and support in preparing this manuscript. Anna Flynn at Arizona State University provided important input and a framework for Chapter 8 on planning. Jackie Wilcock, also, at Arizona State University, provided essential support with word processing and preparation of figures.

Any inaccuracies or omissions are solely the responsibility of the authors.

Lisa M. Ellram
Laura M. Birou

CONTENTS

Series Overview   vii

Preface   ix

Chapter 1: Introduction

The Operational Perspective of Purchasing   1
The Evolution of the Purchasing Function   2
The Role of Purchasing in Total Customer Satisfaction   3
The Strategic Role of Purchasing   3
Organization of This Book   5

Chapter 2: Forecasting

Types and Purposes of Forecasts   7
Forecasting Versus Planning   9
Forecasting Terminology   12
       Economic Indicators   12
       The Concept of Indexes   13
       Price Indexes   15
       Industrial Production Index   17
Data Sources: General Forecast Information   19
       External Data   20
       Government Data   20
       Professional Associations   22
       Commercial Sources   24
       Internal Forecasts   24
Forecasting Methodologies and Techniques   24
       Fundamental Issues   25
       Choosing the Forecast Method   26
       Causal Forecasting Approach   29

Market Research   30
Qualitative Methods   31
Factors That Affect Forecast Accuracy   31
Implications of Forecasting on Procurement   34
Key Points   37
Suggested Readings   38

Chapter 3: Strategic Planning Process

Defining Strategy   41
Hierarchy of Business Strategies   42
    Corporate Strategy   42
    Business Strategy   48
    Functional Strategy   49
Strategic Planning Framework   50
    The Nature of Markets   50
    International Markets   54
Strategic Planning Process   55
    External Analysis   55
    Internal Analysis   56
    General Economic Issues   58
    Issues in International Procurement   61
Key Points   62
Glossary of International Procurement Terms   63
Suggested Readings   64

Chapter 4: Managing Internal Relationships

Interfunctional Relationships   67
    Communication Vehicles   68
    The Role and Perception of Purchasing   72
    Relationships with Other Functional Areas   73
Interdepartmental Interaction Styles   84
    Evolution of Business Structure   84
    Using Teams and Committees   91

Key Points   91
Suggested Readings   92

Chapter 5: Purchasing's Role in External Relations

Benefits of Good Supplier Relationships   93
Means of Promoting good Supplier Relationships   95
Supplier Education, Involvement and Development   98
Supplier Education and Training   98
Supplier Involvement   99
Continuous Improvement   103
Supplier Alliances/Partnerships/Preferred Suppliers   104
Types of Partnerships   105
An Alternative Partnership Classification   107
Benefits and Risks of Alliance Relationships   110
Alliance/Partnership Development   113
Supply Base Reduction/Rationalization   116
Supply Chain Management   117
Issues in Reciprocity   118
Associations   119
The External Role and Perception of the Purchasing Function   123
Key Points   124
Suggested Readings   125

Chapter 6: Computerization and Its Impact on Purchasing

Computer Basics   127
Benefits of Computerization   128
General Computer Uses   129
Basic Computer Applications in Purchasing   132
Advanced Computer Applications in Purchasing   136
EDI   136
Types of EDI Systems   137
EDI Implementation   142
Bar Coding   144

Electronic Catalogues 144
Decision Support Systems 145
Artificial Intelligence 146
Impact of Computerization on the Future of Purchasing 147
Key Points 151
Appendix 148
Basic Overview of Computer Systems 148
Key Points 152
Glossary of Basic Computer Terminology 152
Suggested Readings 154

Chapter 7: Social Responsibility and the Disposal of Hazardous Waste

Social Responsibility 155
Going Green 157
What is Hazardous/Regulated Material? 160
Magnitude of the Waste Management Problem 161
Options for Disposal of Hazardous Waste 161
Legal Issues Regarding Hazardous Waste 165
Ownership and Liability Issues 166
Proactive Strategies for Waste Management 168
Green Purchasing 171
Summary 173
Key Points 174
Suggested Readings 174

Chapter 8: Purchasing Planning and Acquisition Strategy

The Planning Horizon 177
Linking the Three Planning Levels 178
Purchasing and the Planning Process 178
Issues in Purchasing Planning 180
Linking Purchasing Strategies with Forecast 184
Buying Strategies 185

Implementation of Buying Strategies 189
Purchasing Strategy 196
Issues in International Purchasing 200
Financial Considerations 202
Summary 203
Key Points 204
Glossary of International Shipping Document Terms 205
Suggested Readings 205

Index **207**

CHAPTER 1

INTRODUCTION

The purchasing area within many organizations is currently going through many changes. This is often reflected in departmental name changes. The group that used to be called purchasing may now be called sourcing, strategic sourcing, supply management, strategic supply management, supplier management, or materials management, to name a few.

Along with the name changes has come a growing recognition of the importance of purchasing activities to the success of an organization, and a growth and shift in the types of activities performed by the sourcing area.

THE OPERATIONAL PERSPECTIVE OF PURCHASING

Purchasing was once looked upon as primarily a service function. As such, it was the responsibility of purchasing to meet the needs of the requester. It was not the responsibility of purchasing to question those needs, forge long-term relationships with suppliers, or get involved in the needs of the end customer.

This perspective severely limited the contribution that purchasing could make to the firm. In this scenario, purchasers had to focus primarily on a narrow set of activities to serve the needs of the internal interfaces, such as production, marketing, operations, and others who needed to procure something from outside the organization. The scope of purchasing activities was defined and limited by those internal customers. Purchasing focused on getting the right product or service to the right place at the right time—in the right quantity, in the right condition/quality, and from the right supplier at the right price. While this may sound like a broad range of activities, in reality it was not, because the internal customer was defining what was "right" at each step.

This is not to imply that, in this role, purchasing was not making an important contribution to the organization. Purchasing played a key role in keeping the operation running by ensuring a reliable source of supply. In many cases, based on the requests of internal customers or their own initiative, purchasing contributed directly to the bottom line of the organization by reducing prices paid to suppliers. However, this operational perspective focuses on short-term, day-to-day purchasing details rather than the big picture, or systems approach, regarding how purchasing can support the organization's broader goals.

Purchasing was often seen as an activity that could be performed by anyone within the firm. After all, it involved following a series of prescribed steps: writing up a purchase order, contacting suppliers for pricing, and in some cases following up if a supplier did not deliver. Purchasing was seen as a function without specific goals of its own.

THE EVOLUTION OF THE PURCHASING FUNCTION

The purchasing function has gradually evolved. As organizations increasingly automate and outsource many activities, the amount of funds spent on external purchases as opposed to labor increases. Thus, purchasing activities have been receiving more attention. For the past 30 years, and in some cases even longer, enlightened organizations have given purchasing more leeway in performing its duties. In some cases, purchasers have taken the initiative in broadening their roles, in order to contribute more fully to the organization.

In many ways, purchasing today stands at a crossroad in its evolution. Many activities that were once the mainstream bulk of purchasing are being eliminated and automated. Activities such as purchase order placement, expediting, matching documents, and calling to check stock have either been eliminated altogether, or are now possible on-line with Electronic Data Interchange (EDI). While elimination of such routine clerical activities presents an opportunity for purchasing to play a more proactive role in the firm, purchasers must recognize and seize the opportunity, or they may face job elimination.

An important part of recognizing opportunities to contribute comes from understanding the organization's strategic goals and direction, so that purchasing can support those goals. It also comes from understanding the important role that purchasing plays in helping the organization achieve total customer satisfaction.

THE ROLE OF PURCHASING IN TOTAL CUSTOMER SATISFACTION

In the past few years, many successful organizations have changed the fundamental premise by which they operate. Most businesses once said that their purpose was to make a profit for the owners/shareholders. While that is obviously still a goal, many businesses have aligned themselves with government and not-for-profit organizations, which see their goals as "serving the customer" or "providing a service to the customer." This shift in philosophy reflects the fact that many organizations have realized that, if they do not serve the customer effectively by meeting some otherwise unfulfilled need, they will cease to exist.

Traditionally, purchasing has been separated from the firm's end customers. However, high-quality, reliable goods and services on a timely basis at a reasonable cost often directly affects customer satisfaction. The relationship is illustrated in Figure 1-1. An organization cannot provide its ultimate customers with better quality goods and services than it receives from its suppliers. If a supplier is late with a delivery or has quality problems, these factors affect the quality and availability of the product or service to the customer. They also increase the total cost of that product or service.

Thus, it is important that purchasers understand the needs of their organization's customers. Such an understanding will allow purchasing to make the "right" decisions to meet external customers' needs, while also fulfilling the requirements of internal customers.

THE STRATEGIC ROLE OF PURCHASING

The strategic role of purchasing is to perform sourcing-related activities in such a way as to support the overall objectives of the organization. As is discussed extensively in this book, purchasing can make many contributions to the strategic success of the organization. Purchasing plays a key role as one of the organization's boundary spanning functions. As a boundary spanning function, purchasing has many internal and external contacts.

Through external contacts with the supply market, purchasing can gain important information about new technologies, potential new materials or services, new sources of supply, and changes in market conditions. By communicating this competitive intelligence, purchasing can help

FIGURE 1-1
Total Customer Satisfaction Depends on Supplier Performance

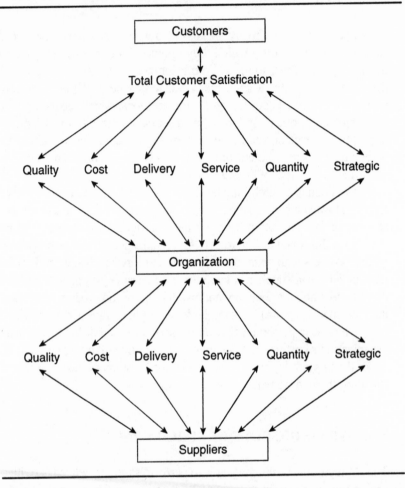

Source: Michiel Leenders and Anna Flynn, *Value-Driven Purchasing: Managing the Key Steps in the Acquisition Process,* Burr Ridge, IL: Richard D. Irwin, Inc., 1995, p. 3.

reshape the organization's strategy to fit within constraints and take advantage of opportunities.

Purchasing can also help support the organization's strategic success by identifying and perhaps even developing new suppliers. Getting suppliers involved early in the development of new product and service offerings, or modifications to existing offerings, can reduce product/service development times. The whole idea of time compression—getting out to market quickly with a new idea—can be very important to the success of that idea, and perhaps to the organization's position as a leader or an innovator.

ORGANIZATION OF THIS BOOK

This book explores purchasing strategy and some of the key issues that have an impact on purchasing strategy. In addition, it examines some of the very important but rapidly changing areas in purchasing, such as computerization and materials disposal.

This book takes a broad, strategic perspective on the role of purchasing in the organization, based on the notion that the purchasing area's contribution to the organization's success has great potential.

The authors believe that purchasing professionals should proactively pursue improvement in their suppliers, in their organizations, and in their own purchasing processes. That belief is reflected throughout this book.

Chapter 2 discusses the need for purchasing forecasts and provides an overview of various forecasting methodologies and terminology. The chapter closes with a discussion of the implications of forecast data on purchasing and the organization's operations.

Chapter 3 introduces the reader to the basic concepts of planning: short-term, intermediate, and long-term planning. It provides a linkage to the forecasting data discussed in Chapter 2, and discusses specific types of forecast data that purchasing may use. The chapter links forecasting with particular types of buying strategies, and it explains various types of contractual agreements and risk reduction strategies. In addition, Chapter 3 discusses the impact of overall economic conditions on the purchasing function. It also includes a Glossary of International Procurement Terms.

Chapter 4 explores the types of internal relationships in which the purchasing function may become involved. It stresses the importance of good communications, mutual respect, and cooperation, and the use of various teams to achieve the organization's overall objectives.

On one level, Chapter 5 is a culmination of the first 4 chapters. It connects purchasing's role in planning and how purchasing fits into the organization with the way that purchasing relates to the external environment. A key discussion in this chapter involves purchasing's relationship with suppliers, including the idea of "partnering" or preferred supplier relationships. This chapter also discusses the important role of purchasing in representing the firm to outside organizations, such as trade and professional associations. A listing of some of the major professional associations with which purchasers are affiliated in the United States is included as a reference.

Chapters 6 and 7 deal with two major topics that are of great concern and are receiving increasing attention in purchasing: computerization and environmental issues. Computerization is a way of doing business today. Chapter 6 discusses the computerization of the basic purchasing process, and provides a definition of some basic computer terminology. The bulk of the chapter discusses the key new information technology that is available, and the profound effect that some of this new technology is likely to have on purchasing operations in the future. As noted in this chapter, most of the routine, repetitive activities once performed by purchasing can be easily and efficiently automated via computers, electronic data interchange, bar coding, and similar technologies. Other technologies, such as artificial intelligence and decision support systems, represent major opportunities for purchasing to improve its decision making abilities. Thus, computerization represents an opportunity for purchasers to move out of clerical activities and instead use their time to proactively support the organization's strategic goals and objectives. An appendix gives a Basic Overview of Computer Systems. A glossary of Basic Computer Terminology is included.

Chapter 7 overviews the increasingly important area of environmental concerns, and the impact that such concerns have on purchasing. Purchasing is increasingly being called upon to take a leadership role in managing materials issues, particularly hazardous materials flows. In organizations that do not directly purchase or create hazardous materials, purchasing can still be held responsible if suppliers generate hazardous materials and create environmental issues in serving the purchaser's organization. This chapter discusses methods of minimizing waste, disposal options, and political and social issues related to hazardous materials.

With an understanding of basic planning and forecasting concepts, as well as how purchasing fits into the overall organization, Chapter 8 discusses the role of purchasing in strategy and policy development. This chapter discusses the scope of the purchasing function's activities, and purchasing's role in recommending strategy in private, public, and not-for-profit organizations. Purchasing's role in product/service design, legal responsibilities, and international purchasing is also discussed.

CHAPTER 2

FORECASTING

Accurate forecasts are a key component of developing an effective, proactive purchasing environment. Forecasts provide the visible basis for planning the efficient, timely flow of goods and services to the organization and its customers, guiding the actions employed by purchasers to meet these needs. Forecasts cover a broad range of environments, from the macro-level, such as expected general economic conditions, to the micro-level, including the expected sales or revenue of the organization and the pricing and availability of supplies. The forecast is the primary tool for matching the external demands of the marketplace with the internal capabilities of the organization.

TYPES AND PURPOSES OF FORECASTS

A forecast is a prediction of a future event or state of affairs. As such, the forecast acts as the catalyst for the entire operating system. It represents a quantitative translation of the overall demand reflected in the business plan into tangible inputs, via the transformation process known as the product-delivery process. It acts as the linkage between the strategic vision and the operating environment (Figure 2-1). The forecast can be viewed as the engine that drives the system. Its importance cannot be overemphasized, as the effectiveness of the entire system is dependent on the quality of the information contained in the forecast.

Just as inventory accuracy is the single most important factor in effectively minimizing stock levels, forecast reliability is the linchpin of any inventory planning system. As more managers are recognizing, the company that can forecast SKU demand more accurately than its competitors

FIGURE 2-1
Forecasting Framework

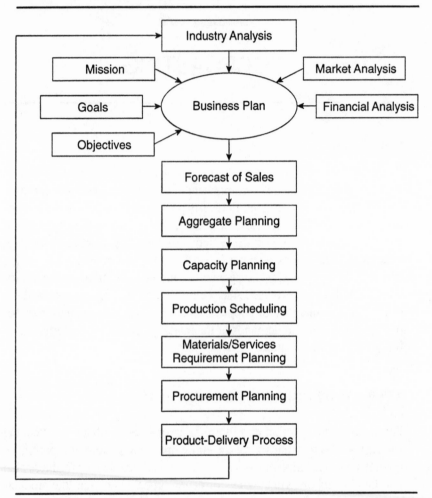

has a huge strategic advantage: it can optimize customer service with minimal capital.[1]

Unlike a simple guess, a forecast is generally based on historical data, cause-and-effect relationships, expert opinion, or a combination of these

[1]"How to Choose the Best Inventory Forecasting Software," *Inventory Reduction Report,* September, 1991, pp. 5-7.

factors. The goal of forecasting is to provide sufficient information for strategic planning purposes. As the definition of forecasting indicates, it is a prediction, not a guarantee of future events. Forecasts are therefore subject to error and inaccuracy. A "good" forecast is able to provide accurate, reliable information, minimizing the degree of uncertainty in the organization's operating environment. This "vision" of the future allows management to plan the effective utilization of resources in support of the corporate strategy.

Without the ability to plan for the future, organizations would be unable to determine their staffing requirements, supply needs, material requirements, capacity, and inventory of finished goods to support the demand from the market, which translates into sales for the company. In the absence of a forecast, decisions regarding future operations would be made in total isolation, resulting in sub-optimal performance of the organization and the functions that support the operations of the organization, including purchasing.

Organizations develop a variety of forecasts internally: forecasts of sales volume based on expected customer demand; production forecasts to support expected sales; supplies and materials required to meet production targets; forecasts on the availability of purchased material; and forecasts of prices for both inputs into the transformation process and outputs in the form of finished products or services. The relationship between the different levels of forecasts is cascading, or a "trickle-down" chain of events. As in Figure 2-1, the most aggregated level, forecast of customer demand in terms of sales, acts as catalyst or input in the determination of the dependent demand forecasts further down the value-added chain.

FORECASTING VERSUS PLANNING

Finally, forecasting should be distinguished from planning. While a forecast is a prediction of the future, planning involves how to act to meet future expectations, or forecasts. The purpose of the forecast is to provide information. The role of planning is to help prepare the operating environment to meet demand in an efficient, cost-effective manner. The role of purchasing is to make the forecast "come alive" through the procurement of the appropriate materials and services to support the anticipated level of demand.

In setting the price and availability of its finished products or services, the organization also needs a great deal of information regarding the pricing and availability of inputs. Thus, the purchasing function may be called upon to make forecasts of supply market conditions. Predictions of supply prices and availability may impact the organization's buying patterns, as well as the pricing of products and services.

> Forecasts do influence the requirements placed on purchasing, but before the orders have been placed on vendors, a number of other factors have entered the situation. These other factors heavily influence the prices charged for the products being manufactured and through these finished product prices, affect profit or loss. A few of these factors are connected with production and operating costs (e.g., taxes, materials costs, wages, overhead, etc.).

> At the same time, such factors may further influence the choice of vendors, methods of transportation, and future order placements. At this point, the linkage between purchasing and forecasting becomes even clearer, for now such global considerations in the economy as recession and expansion, unemployment and inflation, or lower or higher tax rates must be interpreted in terms of the individual company.[2]

Sales forecasts are a prediction of the "independent demand" generated by the interest of the customer in the end item or service. This type of forecast is the most difficult to derive, and is subject to the highest level of error. The disaggregated forecast of customer demand presents itself to the purchasing environment in the form of demand for raw materials, component parts, sub-assemblies, and services. This type of demand is often referred to as "dependent demand" or derived demand, as the quantity of material required by the production system is dependent on how operations schedule and produce what is required.

Independent demand is comprised of four basic components, which demonstrate patterns over time. These components are the average level of demand, trend, seasonal, and cyclical influences. A forecast may incorporate one or all of these elements, depending on the type of product and the environment under consideration. Another element of demand, which does not demonstrate any recognizable patterns over time, is random error. Random error cannot be predicted, and is caused by chance variation in demand. In an accurate forecast, the random error component will be small compared to the other demand characteristics.

The baseline in developing a forecast is generally a history of demand that is averaged over time. For example, purchasing may review previous-year purchases by month. The trend component indicates the

---

[2]Cuelzo, Carl M., PhD. "Purchasing and Forecasting." *St. Louis Purchaser,* March 1990, p. 19.

presence of a continued bias in the demand, indicating growth or decline. If the forecast is not adjusted for this type of systematic bias, the forecast will always be below the actual level of demand if the trend is upward, or always above the actual level of demand if the trend is downward.

Many products and industries are subject to seasonal influences. Some obvious examples of this influence are found in the sporting goods industry, and in the travel and leisure business. Seasonal influences occur annually, while cyclical fluctuations can occur over a much longer period of time. Forecasts for durable goods are more likely to incorporate a cyclical element. The automotive industry has typically demonstrated a five-year cycle, while the housing industry has a much longer cycle of nearly 20 years. The length of these cycles may change over time due to foreign competition, product life cycles, and economic factors.

> The Austad Company, a cataloger of golf equipment, indicated that they adjust the forecast based on the life cycle of the product to improve forecast accuracy. For example, certain models may have a product life cycle of only one season while others may remain in production for many seasons, depending on the product development time and consumer preferences. Thus, they are concerned about seasonal as well as other factors that impact demand.[3]

In the fashion skiwear business, demand is heavily dependent on a variety of factors that are difficult to predict—weather, fashion trends, the economy—and the peak of the retail selling season is only two months long. Even so, Sport Obermeyer has been able to eliminate almost entirely the cost of producing skiwear that customers don't want, and the cost of failing to produce skiwear that they DO want, by using accurate response (a forecasting method.) The company estimates that by implementing this approach, it has increased its profits by between 50 percent and 100 percent over the last three years.[4]

The level of customer demand is affected by factors both internal and external to the operating environment. Internal factors involve those decisions made by management that impact the level or timing of independent demand. The contemporary term used to represent these internal management decisions is "demand management." These decisions can be of a proactive or reactive nature. The most common approach involves the manipulation of

---

[3]Wisner, Joel D. "Forecasting Techniques for Today's Purchaser." *NAPM Insights,* September 1991, p. 23.

[4]Fisher, Marshall L. et al. "Making Supply Meet Demand in an Uncertain World." *Harvard Business Review,* May 1994, p. 87.

price in the form of rebate programs (automobiles, credit cards), off-peak pricing (telephone companies, airlines), bundling products or services (automobiles, frequent flyer programs) or volume discounts. Purchasers must be made aware of such programs so they can anticipate increasing purchases to support the additional production required to meet increased demand.

External factors represent those elements beyond the control of management. These would include the overall state of the economy, government actions, competitors' actions, and changes in customer needs. Since these factors cannot be controlled, the ability to forecast these changes and respond to them faster than competitors is one source of competitive advantage. To facilitate this process, leading, coincident, and lagging economic indicators have been identified for the purpose of forecasting, and will be discussed later in this chapter.

FORECASTING TERMINOLOGY

There are a number of terms that are key in understanding, using, and developing forecast data. This section provides a foundation by discussing the concepts of key economic indicators, indexes, and interest rates. Also discussed are inflation and deflation, industrial production indexes, and the concept of capacity utilization.

Economic Indicators

Economic indicators are activities that change relative to the economy. They are used as a method of forecasting business cycles. The Bureau of Economic Analysis of the U.S. Department of Commerce groups key indicators of economic trends according to how they coincide with changes in the economy: leading, coincident, and lagging. The indicators were initially developed around World War I, and they have been revised on occasion—most recently in 1989—to improve their predictive ability.

Leading indicators precede changes in the general economy and have historically demonstrated up to 9.5 months lead for business cycle peaks, and 4.5 months for business cycle troughs.[5] There are currently 11 leading indicators. The composite index includes the average workweek of production workers in manufacturing, average weekly state unemployment

---

[5]Wright, John W., Editor. *The Universal Almanac.* Kansas City: Andrews and McMeel, 1994, p. 250.

insurance claims, new orders for consumer goods, vendor delivery performance, contracts and orders for plant and equipment, changes in manufacturers' unfilled orders for durable goods, change in sensitive materials prices, index of stock prices for the 500 common stocks, the amount of new money in circulation, and index of consumer expectations. Each of these indexes represent business expectations of future economic conditions. Due to the predictive capability of leading indicators, they are extremely useful in forecasting and planning.

Coincident indicators operate in harmony, or concurrently, with changes in general economic activity. These indicators have very little predictive value, and are therefore of limited use in forecasting business activity for the purposes of proactive planning. The frustration associated with developing accurate forecasts is illustrated by the following quote: "For many forecasters and analysts, the toughest part of trying to predict where the economy will be tomorrow is knowing where it is today."[6] Included in this group of indicators is the index of industrial production, non-agricultural employment, manufacturing and trade sales, and personal income less transfer payments.

Lagging indicators represent those activities that tend to change after the state of the general economy has changed—4.5 months for peaks and 8.5 months for troughs.[7] Indicators are thought to reflect the cost of doing business and general consumer consumption patterns. The Consumer Price Index (CPI), average length of unemployment, ratio of consumer installment credit to personal income, the ratio of manufacturing and trade inventories to sales, the change in the index of labor cost, the average prime interest rate, and commercial and industrial loans outstanding are the lagging indicators of economic activity.

Figure 2-2 provides a summary of the various economic indicators. Data sources for these economic indicators include Dun & Bradstreet, F.W. Dodge Corporation, Dow Jones, Bureau of Labor Statistics, Department of Commerce, and the Federal Reserve Board.

The Concept of Indexes

Simply defined, indexes are factors that are used to adjust numerical data so the data can be compared between different periods of time. Indexing

---

[6]Keen, Howard, Jr. "Use of Weekly and Other Monthly Data as Predictors of the Industrial Production Index." *Business Economics*, January 1988, p. 44.

[7]Wright, p. 251.

FIGURE 2-2
Economic Indicators

Economic Indicators	
Leading Indicators	Average Manufacturing Workweek
	Average Weekly State Unemployment claims
	New Orders for Consumer Goods
	Vendor Delivery Performance
	Contracts and Orders for Plant and Equipment
	Change in Unfilled Orders for Durable Goods
	Change in Sensitive Materials Prices
	New Money in Circulation
	Index of Consumer Expectations
	Index of 500 Common Stocks
Coincident Indicators	Index of Industrial Production
	Non-Agricultural Unemployment
	Manufacturing and Trade Sales
	Personal Income Less Transfer Payments
Lagging Indicators	Consumer Price Index
	Average Length of Unemployment
	Ratio of Installment Credit to Personal Income
	Ratio of Manufacturing and Trade Inventories to Sales
	Change in Labor Cost Index
	Average Prime Interest Rate
	Commercial and Industrial Loans Outstanding

starts with a base year, and a value of 100 is assigned to the base year. Comparisons between time periods are then given as a percent change from the base year. In using and interpreting indexes, it is important to know what year the base period represents, as it is the benchmark for comparison in constant dollars.

For example, the Producer Price Index (PPI) uses 1982 as the base year for comparative purposes. The value of the PPI for 1982 is equal to 100. To compare price changes from one period to another, the index for the ending year is divided by the index from the beginning year. This value is subtracted from one, then multiplied by 100 to determine the percent change from the base year. For example, in comparing the base year of 1982, which is equal to 100, and 1985, which has an index of 105, the percent change from 1982 to 1985 is five percent:

$$[105/100 - 1.0] \times 100 = 5\%$$

The value of utilizing a base year for comparative purposes is demonstrated when evaluating two time periods that do not include the base year, as the index provides information in constant dollars. For example, if 1993 had an index number of 130, then comparing it to 1985, which has an index of 105, would yield the following:

$$[130/105 - 1.0] \times 100 = [1.238 - 1.0] \times 100 = 23.8\%$$

The conclusion drawn is that 1993 is 23.8% higher than 1985 in constant dollars.

Price Indexes

The three most common indexes used by purchasing are Producer Price Index (PPI), Consumer Price Index (CPI), and the implicit price deflator (GDP deflator). These are all broad indicators of prices between a base year and another year.

The Producer Price Index (PPI) is reported as an "all commodities" index, and prior to 1978, it was referred to as the Wholesale Price Index (WPI). It measures the average change in prices paid on thousands of items in the manufacturing and mining sectors of the economy at various stages of processing. The data is collected from sellers involved in the first significant large-volume commercial transactions for the commodity of interest.[8] The information is compiled monthly by the U.S. Department of Labor's Bureau of Labor Statistics from nearly 500 mining and manufacturing organizations. This information is widely available in the business press and from a number of government sources, including a monthly periodical entitled *Producer Price Index*. This broad index is a composite of 8,000 indexes for specific product categories and over 3,000 commodity price indexes. The aggregate measures are decomposed by stage of processing, including finished goods, intermediate materials, supplies and components, and raw materials. This disaggregation of the PPI can provide a more accurate gauge of price trends for a specific type of buy. To further aid in accuracy, the PPI is seasonally adjusted. The PPI is often used in contracts as a basis for price escalation clauses.

The PPI is also commonly used in negotiation. According to Maurine Haver of Haver Analytics, "When a contract based on fixed costs is up for renewal, a purchaser may refer to the PPI to determine what the

---

[8]Wright, p. 253.

supplier will be asking for . . . you can reference the PPI to see if these commodities increased in price. Then, if the supplier requests an increase, you will know if industry costs have increased."[9] Another example of utilizing the PPI in negotiations is provided by the John Deere Company.

> A major supplier of computer paper notified John Deere of a price increase to go into effect October 1, due to price increases by this company's suppliers of rolled paper. The increase was to be from 3.0% to 7.5% for the various papers they supply.
>
> John Deere Waterloo Works purchasers reviewed the relevant PPI retroactive to November, 1984, when the supplier started providing material. This was compared to prices paid to the supplier for the same time frame.
>
> A comparison of the data revealed that the price increases were inconsistent with the market information provided by the PPI for raw material decreases, which the supplier had not passed on to the buying organization. This analysis resulted in no price increases incurred for John Deere, which would have amounted to approximately $1,300 a month.[10]

The Consumer Price Index (CPI) is a common measure of inflation. It represents the change in prices for a fixed set of household goods and services bought by the average consumer on a regular basis and referred to as a "basket" of goods. The specific goods and services that compose this basket are reviewed approximately every ten years, and are adjusted to represent consumer buying patterns. The CPI is also referred to as the "cost of living index," as it indicates the seasonally adjusted, relative buying power of personal income compared to the base period of 1982 to 1984.

The Bureau of Labor Statistics spends about $26 million annually to collect price data, including direct taxes, from 57,000 households and 19,000 establishments in 85 areas across the country.[11] This information is compiled and released in two forms. The CPI-U represents Urban Consumers (80 percent of the population), including wage earners and clerical workers, professional, managerial, and technical workers, the self-employed, retirees, short-term workers, and the unemployed. There is also the CPI-W for Urban Wage Earners and Clerical Workers, which represents approximately 32 percent of the population.[12]

---

[9] Budding, Gonad. "Applying the PPI to Purchasing." *NAPM Insights,* February 1991, p. 6.

[10] Budding, p. 6.

[11] Wright, p. 253.

[12] Wright, p. 253.

The Gross Domestic Product (GDP) deflator, formerly referred to as the Gross National Product (GNP) deflator, was changed to reflect the exclusion of net property income that is generated abroad. GDP measures the total national output of goods and services valued at market prices at the final point of consumption. The Department of Commerce utilizes the Implicit Price Deflator to assess the value of current production in current prices for the same goods and services for prices in the base year of 1987.

The Implicit Price Deflator is calculated by dividing the current value of GDP by the constant value GDP (1987) and multiplying the ratio by 100. In 1992 the GDP was valued at $5,950.7 billion in current dollars, while the 1987 constant value GDP was $4,922.6 billion. The Implicit Price Deflator is therefore, [$5,950.7/$4,922.6] x 100, or 120.9. It represents the comparison between 1992 and 1987 prices.[13]

Industrial Production Index

Around the middle of every month the Federal Reserve Board releases the Industrial Production Index (IPI) for the previous month. The IPI measures the level of output in manufacturing, mining, and electric and gas utilities. This seasonally adjusted index differs from other indexes in that it measures the volume of output in units, not the market value of those units. Unlike the Gross Domestic Product (GDP), the IPI is a summation of total output, including exports. The IPI uses 1977 as its base year for comparative purposes.

> The keen interest in getting a better handle on the current state of the economy pertains in no trivial way to the monthly release of the Federal Reserve Board's Industrial Production Index (IPI). Whether the interest lies in the performance of the industrial sector directly, in getting a reading on the reasonableness of a near-term forecast of Industrial Production, or primarily in the reaction to the reported performance in the industrial sector, the next monthly IPI release offers an important clue as to which direction output in this sector is heading.[14]

Inflation

Inflation is the measure of sustained average price increases in an economy. It is brought about by the natural market forces of supply and demand,

---

[13]Wright, 1994, p. 253.
[14]Keen, 1988, p. 44.

and it is described as "too much money chasing too few goods." There are actually two generic causes of inflation. The first is referred to as cost-push inflation, which is a result of increases in the cost of inputs or resources in the production of goods and services during the product-delivery process. The second cause, demand-pull inflation, results when sustained demand in the economy is higher than the full employment level.[15]

The rate of inflation has generally ranged from three to nine percent annually since the 1940s, with peak inflationary periods occurring during the mid-to-late 1970s. The United States economy has not realized a deflationary period since the 1930s. Buying patterns related to the timing of major purchases are significantly influenced by the rate of inflation. As the inflation rate rises, there is an incentive to purchase immediately in order to maximize the buying power of the dollar. However, the cost of carrying inventory must also be factored into the purchase decision.

Assumptions about inflation and deflation are important in forecasting future costs and prices. The inflation/deflation rate is generally expressed as a percentage change in prices relative to a baseline. People are generally concerned with price changes from one year to the next, because the changes influence the overall purchasing power of the unit of currency.

In some countries where inflation is considered "out of control," the procurement strategy that predominates involves the acquisition of as much material as possible under current market conditions. Such "forward buying" deals with the rapid decline in the purchasing power of the local currency in comparison to the cost of carrying inventory.

Clauses addressing the impact of inflation and deflation on the negotiated price are often incorporated in long-term contracts and partnering relationships in order to protect both the buyer and the supplier from risk. These clauses cover unexpected and sustained increases or decreases in the cost of inputs, including labor, raw materials, and component parts.

Interest Rates
The supply of money in the economy, together with the corresponding level of demand for the money, serve to determine the interest rate—or the cost of borrowing money. The Federal Reserve Board utilizes the discount rate, which is the rate it charges commercial banks to borrow

---

[15]Wright, p. 252.

money, to manage the growth rate of the economy and the rate of inflation. The discount rate influences the interest rate that commercial lending institutions charge consumers to borrow money. Interest rates are also affected by the perceived level of risk, length of the repayment period, and the expected rate of inflation.

Higher interest rates encourage the repayment of debt and promote savings. They are designed to slow the growth and inflation rate in the economy. The reverse is also true. Lower interest rates encourage borrowing and spending, which serve to spur growth in the economy. Interest rates are significant in forecasting because they themselves need to be forecasted. They represent one perception of future business conditions.

Commonly referenced interest rates include the Federal Reserve discount rate and the Prime rate. The latter represents the rate of interest that banks charge their "best" customers when loaning money. Three-month U.S. Treasury Bills, U.S. Treasury Notes and Bonds, and AAA Corporate Bonds are also commonly quoted interest rates. Information regarding these interest rates is available in *The Wall Street Journal,* or in the business section of most daily papers.

Capacity Utilization

Another important piece of data provided by the Federal Reserve System is the level of capacity utilization. This is calculated by dividing the total output determined in the Industrial Production Index (IPI) by the level of available capacity in the represented industries (manufacturing, mining, and electric and gas utilities). This information is valuable in evaluating the relative health of the industries and the economy. High utilization rates indicate an active economy, effective asset management, profitability, and overall health of the respective industries.

DATA SOURCES: GENERAL FORECAST INFORMATION

In developing forecasts, there are many sources of data available to the organization and the purchasing function. External data sources include federal and local governments, private publications, regional forecasts, and commercial forecasts that may be specifically commissioned or purchased by the organization. In addition, forecasts of economic, competitive, and social conditions that may affect the organization can be developed by internal experts.

External Data

External data can be relatively inexpensive or free to obtain. Such data may also be available quickly and easily. In addition, external data may tap sources that are not available to the organization, such as U.S. manufacturers' census information. External forecast data may also be developed by experts whose knowledge and capabilities in the forecast area outstrip those of personnel within the organization.

However, external data is not without drawbacks. External data is often outdated. For instance, census data is only collected every 10 years. In addition, the data may not be presented in a form that is readily usable by the organization. Further, some external data is inaccessible or expensive to purchase. The advantages and disadvantages of external data will be discussed further as we examine the major types of external data.

Government Data

The primary source of external data for forecasting purposes comes from local, state, and federal government agencies. The majority of this information is available free-of-charge, or at a minimal cost. The largest problem lies in uncovering what is readily available and determining the appropriateness and/or usefulness of the information. Often, the data are too old to be of value in forecasting. On-line data searches have improved the efficiency of the information search process. In addition, telephone inquiries addressed to research librarians, information personnel at the Library of Congress, or specific government agency representatives can be beneficial. Many government agencies publish a listing of the reports they generate each year, including a brief description of their content.

In addition to the publications provided by the Bureau of Labor Statistics of the U.S. Department of Labor, the Department of Commerce, and the Federal Reserve Board, purchasing and materials management professionals should also consult the *Survey of Current Business,* the *Federal Reserve Bulletin,* the *Statistical Abstract of the United States,* the *Economic Report of the President,* and *Business Conditions Digest.* What follows is a brief discussion of the information available in each of these publications.

Survey of Current Business
The Bureau of Economic Analysis (BEA) of the U.S. Department of Commerce is responsible for the publication of the *Survey of Current Business.* This monthly publication presents a snapshot of the current

business situation for both the public and private sectors. It includes valuable information on business cycle indicators, business statistics, corporate profits, national income, Gross Domestic Product, and foreign direct investment, among other topics.

Federal Reserve Bulletin
In addition to the monthly calculation of the Industrial Production Index (IPI) and Capacity Utilization, the Federal Reserve Board also publishes the *Federal Reserve Bulletin*. This publication contains information on monetary policy, significant legal developments, and financial and business statistics related to interest rates, savings, wages, and prices for both the domestic and international arenas.

Statistical Abstract of the United States
Published by the Bureau of the Census of the U.S. Department of Commerce, the *Statistical Abstract of the United States* contains all of the "vital statistics" for the U.S. economy. Information pertaining to the demographics of the U.S. is provided, such as population, income, expenditures, disposable income, and wealth.

Some of the more valuable information contained here for the purchasing and materials management professional includes the Gross Domestic Product, prices, purchasing power, cost of living index, interest rates, and vital statistics on specific industries including communications, energy, transportation, agriculture, mining/minerals, construction, and manufacturing. The document also contains comparative international statistics on prices, imports, exports, and international transactions.

Economic Report of the President
Issued to Congress, this annual report represents an overview of the current economic situation, economic policy, and the agenda for the upcoming year. It is basically the administration's forecast of overall economic activity. Included in this report are relevant monetary and fiscal policy, employment and productivity trends, regulatory reforms that affect markets and industries, the expected growth rate, productive capacity, and an overview of international policy (exchange rates, trade agreements, etc.). This may be particularly useful to purchasers who are buying internationally.

Business Conditions Digest
Business Conditions Digest provides updated information on cyclical business indicators, including the leading, coincident, and lagging composite indexes. In addition, cyclical indicators by economic process, diffusion

indexes, and rates of change are included in this report. Other important economic measures that forecasters can garner from this publication include prices, wages, productivity, employment, industrial production, and international activity.

Data from these government agencies is available on the aggregate level or on a disaggregated basis by specific location (state, region, etc.) or SIC code. A significant portion of this information would be relevant to a comprehensive commodities study.

Additional international data is provided in annual reports generated by the United Nations, such as "International Trade Statistics," "World Economic Survey," and "Statistics of World Trade in Steel."

Professional Associations

Another source of external data is provided by professional associations such as the National Association of Purchasing Management (NAPM), the American Production and Inventory Control Society (APICS), and industry and trade associations on the regional and national level.

The NAPM Business Survey has been collecting information on business activity since the 1940s reporting the results of their analysis in the NAPM *Report on Business* (ROB), which is published the first Monday of every month for the preceding month. This information is widely accepted as a valuable forecasting tool and is published in major newspapers, business periodicals, and financial newspapers. It is also reported on the major news networks.

The ROB surveys over 300 purchasing executives representing over 20 manufacturing industries. The sample is designed to be geographically dispersed, and the representation is weighted by the Standard Industrial Classification (SIC) code in proportion to the industries' contribution to GNP. The monthly questionnaire asks the respondent to indicate the relative change in activity compared to the previous month (positive/higher, same/no change, negative/lower) in the areas of production, new orders from customers, purchased inventories, employment, supplier delivery, commodity prices, new export orders, and imports. New orders and imports were added to the survey in 1988 and 1989 respectively.[16] The data for each of these variables is then transformed into a "diffusion index" by adding all

---

[16]Bretz, Robert S. "Forecasting with Report on Business." *NAPM Insights*, August 1990, p. 22.

of the positive/higher responses with half of the same/no change responses. This information is then seasonally adjusted with factors provided by the U.S. Department of Commerce.

Interpreting NAPM's diffusion index is fairly straightforward. The index can range in value from 0 to 100, though these extreme values are rarely realized. A value of 50 indicates that there is no change in the activity level compared to the previous month. A value greater than 50 indicates there is expansion or growth in the activity level, while a value less than 50 indicates that there is a contraction or decline in the activity level. A diffusion index differs from other indexes in that it reports the direction of change in activity level, *not* the absolute level of activity or the magnitude of change.[17]

In 1979 five of these individual measures—new orders from customers, production, supplier delivery, purchased inventories, and employment—were combined into a composite index known as the Purchasing Managers Index (PMI). This aggregated index is probably the most familiar to the purchasing community, and it is widely publicized in the popular press.

Four of the five variables utilized in the PMI composite index demonstrate a positive relationship with changes in business activity. These include new orders from customers, production, purchased inventories, and employment. Supplier delivery, however, actually demonstrates a negative or inverse relationship with changes in business activity. As business activity expands, suppliers usually require longer lead times to meet demand.

Since the ROB and PMI focus on the manufacturing sector of the overall economy, the diffusion index utilizes the standard value of 50 as the centerline (which indicates no change) when forecasting changes in the manufacturing industries. When utilizing these instruments to predict changes in the overall economy, which includes the service sector, the centerline is adjusted down to a value of 44, in order to compensate for the lack of representation of the service sector in the survey.

The primary advantage of utilizing a diffusion index is that it functions as a leading indicator. Historically, the PMI has predicted changes in the business cycle every turn by three months since 1948.[18] Two other advantages of the index are the timeliness and accuracy of the information.

---

[17]Klein, Philip A. and Geoffrey H. Moore. "NAPM Business Survey Data: Their Value as Leading Indicators." *Journal of Purchasing and Materials Management,* Winter 1988, p. 33; and Torda, Theodore S. "Purchasing Management Index Provides Early Clue on Turning Points." *Business America,* June 24, 1985, p. 13.

[18]Klein and Moore, p. 33.

This data is the first available every month and is subject to few, if any, revisions. Compare this to government data, which typically suffers a two-week to several-year delay in reporting, and is still subject to significant revisions.

Commercial Sources

Consulting organizations, market research organizations, industry groups, banks, economic analysis organizations, and specialty organizations are among the other external resources willing to generate forecast information. The primary difference between commercial sources and others is the degree of customization and the corresponding cost. Private sources are willing to customize the forecasting requirements based on the individual customer's needs, but the product is more costly.

Internal Forecasts

Internal forecasts are organization-specific, and they tend to rely on the availability of historical data. The process of generating internal forecasts is initially time-consuming, requiring the identification of accessible and relevant archival data, constructing a database, inputting the historical information, and designing and generating meaningful forecasting reports. The inherent difficulty in forecasting lies in accessing the pertinent information and transforming it into a useful database. The time-consuming and costly nature of implementing an internal forecasting system lends credence to the benefits of requiring a common database utilized by all departments to facilitate the transfer of information.

Historically, organization-specific information of interest to the purchasing function may include sales volume or revenue, lead times, overhead, material, and labor costs. Trend analysis should be utilized to evaluate all of these important elements, and analyzing historical patterns of demand will identify seasonal and cyclical components.

FORECASTING METHODOLOGIES AND TECHNIQUES

This section is devoted to discussion of the various quantitative and qualitative techniques utilized in developing a forecast. The discussion opens with coverage of forecasting environments and the suitability of various forecasting techniques given the environmental parameters. This information is important for purchasing professionals who must determine which technique is appropriate for their organization operating environment.

Fundamental Issues

Key considerations in forecasting include the time period covered by the forecast, the aggregation level of the forecast analysis, the nature of environment being forecasted, the choice of forecasting technique, and the availability of data.

Forecasting time horizons are separated into short-term, which covers up to one year in the future; medium-term for a one- to three-year time frame; and long-term forecasts, which extend beyond three years. As the time frame for the forecast grows, there is a corresponding decline in the quality and an increase in the uncertainty of the information provided by the forecast. On a personal level, it would be similar to detailing your daily activities for tomorrow versus two years from now. You can provide a much higher level of detail with conviction for tomorrow's activities.

Defining the level of the forecast is the next managerial decision that must be made prior to selecting the forecasting technique and data source. Defining the level is important in any business that produces more than one product. The level represents the degree of aggregation of the end products, which are the subject of the forecast. Products can be forecasted by individual end items, product families, or product groups. Independent demand is often forecasted on the highest level of aggregation—product groups—then disaggregated into product families and finally into the demand for individual end items.

The next step in establishing a forecast is an accurate assessment of the nature of the environment that is being evaluated. Included in this assessment is understanding the pattern of demand. To accurately assess the pattern, the following questions must be answered:

- Is the demand extremely volatile, or is it constant?
- Is the demand growing or declining rapidly, or is it relatively stable?
- Is the end item a tangible good that can be inventoried, or must the demand be filled when needed, which is the case in many service industries?
- Is the demand pattern affected by seasonal or cyclical factors?
- Is there an apparent trend in the demand pattern?

Once these questions are answered, it is possible to identify the forecasting technique that will yield the most accurate forecast.

Choosing the Forecast Method

The appropriate forecasting method is determined by matching the length of the time horizons, the nature of the environment, and the availability of data. The four generic methods available are: time series analysis, which is based on the use of historical data; causal modeling, which makes use of regression analysis; market research; and qualitative techniques, which are used in the absence of objective data (see Figure 2-3). The general "rules of thumb" suggest utilizing time series techniques for short-term forecasting, time series or causal modeling for the medium-term, and causal and qualitative methods for the long-term (see Figure 2-4).

Time Series Analysis
There are several common types of time series analysis used to create forecasts. All are based on the use of historical demand to project future demand. The assumption is made that past usage, sales, or prices can be used to predict the future. Moving average, weighted moving average, and exponential smoothing are the three most common forecasting tools.

Moving Average
The simplest method is called the moving average method. The calculation determines the average demand by summing a fixed number of

FIGURE 2-3
Classification of Forecasting Methods

Formal Forecasting Methods	Components
Time Series Analysis	Moving average Weighted moving average Exponential smoothing
Causal Modeling	Multiple regression Correlation models Leading indicators
Market Research	Market Testing Consumer market survey Industrial market survey
Qualitative Techniques	Delphi Expert opinion Sales force composite

FIGURE 2-4
Forecasting Time Horizons

Time Span	Purpose	Forecasting Components
Long Range (*three years or more*)	Capital budgets Plant location Product selection	Delphi Expert opinion Sales force composite
Intermediate Range (*one to three years*)	Budgeting Production/capacity planning Sales planning	Regression Time series analysis
Short Range (*less than one year*)	Production scheduling Inventory control Purchasing Sales forecasting	Trend projection Moving average Exponential smoothing

Source: Markland and Sweigart, "How to Choose the Best Inventory Forecasting Software," *Inventory Reduction Report*, September, 1991, p. 7.

periods of demand and dividing that sum by the number of periods (n). For example, to forecast demand for the next period, the following formula would be utilized:

$$\text{Forecast}_{n+1} = \frac{\text{Sum of actual demand for n periods}}{n}$$

The forecast is updated by keeping the same number of periods but adding the actual demand data as it becomes known, dropping the oldest period's data from the calculation. This is also referred to as a rolling forecast.

The data periods can be days, weeks, months, or years, depending on the level of aggregation needed by the forecast. Selecting the number of demand periods included in the calculation should be governed by the nature of the demand pattern. The forecaster should utilize a lower number of demand periods in an unstable, volatile demand environment, so the forecast reacts quickly to changing patterns of demand. In environments with a constant rate of stable demand, the forecaster can utilize more data periods in the calculation to smooth out minor fluctuations.

Weighted Moving Average
An elaboration of this basic technique is the weighted moving average method. This method allows the forecaster to prioritize the demand periods (D), by weighting their relative contribution to the forecast, shown as W. This

differs from the simple moving average method, which treats all the data periods as equally important in the calculation of the next period's demand.

$$\text{Forecast}_{n+1} = \frac{W_1D_1 + W_2D_2 + W_3D_3 + \dots W_nD_n}{n}$$

The weights are allocated as a percentage, and must therefore sum up to one for the calculation to be accurate. The most common approach involves giving the most recent demand period the highest weight and the oldest data the lowest weight. This procedure ensures that the most current data has the highest level of impact on the next period's forecast. Giving higher weights and priority to current data will cause the forecast to react quickly to changes in the demand pattern. This scenario is desirable under conditions of rapid demand growth or decline.

Exponential Smoothing

The most commonly used time series method is the exponential smoothing method. This method is preferred because it is simple to use, it requires a small amount of data, and it can incorporate trends. The underlying logic recognizes the value of the most recent data in forecasting the immediate future, and weights that information with a smoothing parameter often referred to as alpha (a). Alpha ranges in value from 0 to 1, with larger values placing more emphasis on current data. Therefore, the larger the alpha value, the more responsive the forecast will be to recent changes in demand. The formula for this method is:

$$\text{Forecast}_{n+1} = aD_n + (1 - a)F_{n-1}$$

Where:

a	= a smoothing parameter with a value between 0 and 1
D_n	= the demand for the current period
F_{n-1}	= the forecast for the last period

The average demand included in this calculation, F_{n-1}, can be the product of many periods of historical data. The determination of the number of periods is left up to the forecaster.

To incorporate a trend like the aging population in the forecast, another smoothing parameter called beta (b) is utilized. Beta serves to weight the impact of the trend component on demand.

It is important to remember that one can recognize the presence of a trend even if it is not incorporated in the forecasting technique, because a systematic bias will be evident. The forecast that is not corrected for the presence of a trend will demonstrate an error component, which is: 1) always negative, indicating a declining trend is present; or 2) always positive, indicating an increasing trend.

One measure of the overall accuracy of a forecast is the cumulative sum of the individual forecast errors by time period. The forecast error (E_t), is calculated by subtracting the forecast, designated by F_t, from the actual level of demand, designated by D_t. This yields the equation $E_t = D_t - F_t$. If there is no systematic bias in the forecast, the error component will fluctuate between positive and negative. The desired result is that the cumulative sum of the forecast error (CFE) will equal zero.

There are many other methods of tracking the accuracy of the forecasting technique, including mean squared error, standard deviation of forecast errors, mean absolute deviation of forecast errors, and the mean absolute percent error. All of these methods serve to evaluate the severity of the forecast error, and a severe error indicates that the forecasting technique is not operating properly.

Causal Forecasting Approach

Causal forecasting also utilizes historical data. The data is utilized to identify cause-and-effect relationships between important variables in the establishment of a forecast. The forecast is viewed as a function of internal and external factors that are known to effect the level of demand. Demand is treated as the dependent variable, and the internal and external factors are the independent variables in a linear equation.

Linear Regression
Linear regression techniques identify the equation for a line that "best fits" the historical data. This is done by minimizing the deviation of the data points from the line, also known as ordinary least squares. In simple linear regression, there is one independent and one dependent variable. The equation for the line is $Y = mx + b$. M represents the slope of the line, and b is the point at which the line crosses the Y-axis. The slope of the line (m) demonstrates the direction, positive or negative, and the predictive ability of the independent variable to determine the level of demand. A value of positive or negative one indicates perfect predictability.

Multiple Regression
Multiple regression analysis uses the same logic as linear regression, but incorporates more than one independent variable in the determination of

demand. The variables might include income, population density, population growth, education, geographic location, and others. The correlation coefficient (r) is a measure of the strength and direction of the relationship between the independent and dependent variables. It ranges in value between positive one (both increase together), and negative one (both decrease together). A value of zero means there is no relationship between the variables.

A measure of the accuracy of the forecast is the squared value of the correlation coefficient (r), which is known as the coefficient of determination (r^2). It ranges in value from zero to one, and it indicates the amount of variation in demand explained by the independent variables incorporated in the regression equation. A value of one would indicate that the incorporated variables are perfect predictors of the level of demand.

There are numerous statistical packages available to handle causal forecasting techniques, including the Statistical Package for the Social Sciences (SPSS), Statistical Analysis System (SAS), and Minitab. These software packages are available for both mainframe and personal computers. In addition to regression analysis, these packages can supply a wealth of descriptive information for comparisons of the central tendency of the data. Measures of central tendency include the mean, mode, and median. The mean is the average value of all of the data. The mode is the most frequent value in the data set. The median is the midpoint in the data set, with 50 percent of the values above and 50 percent below the median value.

Range, Variance, and Standard Deviation
Analysis of the dispersion of the data, known as variability, is an important indicator of the degree to which the values within a data set differ from each other. The lower the variability, the more tightly clustered the data is. Measures of variability include the range, variance, and standard deviation. The range is the distance between the largest and smallest value in the data set. The variance is equal to the sum of the differences between each data point in the set and the mean, divided by the number of data points in the set. The standard deviation is calculated by taking the square root of the variance. All of these calculations are measures of the dispersion of the data, and are used to evaluate the quality of the forecast.

Market Research

Conducting market research involves identifying the target population relevant to the research. It is a means of reaching members of the population and gaining their interest in the research effort. The researcher must establish what information is sought from the target population and the method utilized to collect the information. The most common method for gathering large volumes of data is the mail-distributed survey. Another method that is growing in popularity, due to the lack of sufficient responses from mail surveys and the ability it offers to interact with the respondent, is the telephone survey. Live interviews provide the highest level of verbal and nonverbal information to the forecaster, but are very costly to conduct.

Qualitative Methods

Qualitative methods are utilized in the absence of objective data for the purpose of forecasting the probability of an event. Occasions when this situation may arise include the projected sales forecast of a new product, expected changes in consumer preferences, and future technological developments. When no valid historical data exists to support a forecast, the forecaster must rely on expert opinion.

A method widely utilized in the insurance and high-tech industries is the Delphi method. This method relies on the information provided by a group of experts. The process begins by identifying individuals who are considered experts in the subject area. The experts are individually asked to respond to a series of relevant questions. The information is compiled by the researcher, who then distributes the aggregate information back to the pool of experts for further comments and revisions. This process is repeated with strict anonymity until a consensus is reached by the group, usually requiring three to four rounds. The purpose of seeking the anonymous contributions is to eliminate some of the detrimental outcomes of group behavior, including "group think," suppressed creativity, and the domination of the group by an individual member.

In the absence of the time or the resources necessary to conduct market research or utilize a Delphi panel, the forecaster must rely on sales force estimates or executive judgment. This method is often referred to as an "educated guess." While this may not be the desirable course, it is sometimes the only option available.

Factors That Affect Forecast Accuracy

Forecast accuracy is dependent on the applicability of the forecasting method utilized, the accuracy of the input data, and the stability of the planning environment during the duration of the forecast period. Previous sections of this chapter have addressed how to select the appropriate forecasting technique and the importance of data integrity. This section is devoted to the identification of factors that can disrupt the planning environment and weaken the predictive ability of any forecast.

Factors known to affect the accuracy of forecasts include unpredictable events such as war and threats of war, strikes and threats of strikes, natural disasters, political instability, technological changes, and abrupt changes in consumer demand. Inaccurate forecasts result in higher operating costs due to higher inventories, cancellation penalties, material shortages, premium freight, impacts on customer service, lost sales, erratic production scheduling, and the inability to benefit from volume discounts.

Lead Times

In the purchasing environment, fluctuating lead times caused by changes in production schedules and supplier delivery schedules create uncertainty in the planning environment. The primary weapon utilized to guard against this uncertainty and the possibility of material shortages is inventory. An organization's financial picture, production problems, changes in demand, and changing supply market conditions all contribute to lead time fluctuations. Incorporating these issues into a forecast is difficult, as they are often short-term and sporadic in nature. Chronic issues of this sort should be evaluated and included in the forecasting methodology. Purchasing must consider these issues when determining the timing and placement of orders.

> Similarly, quick response and just-in-time (JIT) address only part of the overall picture. A manufacturer might hope to be fast enough to produce in direct response to demand, virtually eliminating the need for demand. But in many industries, sales of volatile products tend to occur in a concentrated season, which means that a manufacturer would need an unjustifiably large capacity to be able to make goods in response to actual demand. For example, Dell Computer Corporation developed the capability to assemble personal computers quickly in response to customers' orders but found that ability constrained by component suppliers' long lead times.[19]

Labor

Labor shortages brought about by the lack of qualified labor or by strikes and threats of strikes can severely impact the forecast. An entire industry can be affected by striking workers who shut down the source of supply.

---

[19] Fisher et al, p. 84.

In the case of strikes in the transportation industries, such as the recent Teamsters strike of United Parcel Service (UPS), the entire distribution chain may be dramatically altered, inflicting material shortages on a broad array of industries.

Forecasting these potential disruptions is possible by staying abreast of upcoming labor contract expirations and renegotiations. The NAPM *Report on Business* (ROB) publicizes the expiration dates of major labor agreements. Armed with this valuable information, the purchasing and materials management professional can assume a proactive posture to cope with the potential supply disruption. Some alternatives include stockpiling inventory, identifying substitute materials, or locating alternative sources of supply.

Labor shortages brought about by an insufficient supply of skilled or trained workers may be local, regional, national, or international in scope. Each of these scenarios presents different problems and potential solutions for the procurement professional. Local or regional shortages can be addressed in the short-term through alternative labor sources from a different region. While utilizing the alternative sources, the proactive purchasing department can work to foster the appropriate skills in local and regional labor sources.

Technology

Shortages on the national and international scope may be a product of changes in the supply and demand conditions for labor, such as the impact of changing technology. This type of situation may require the development of long-term strategic partnering arrangements with suppliers, or the internal development of the skills or abilities needed to make the product. Imbalances in the structural requirements of labor tend to rectify themselves in the long run, due to the natural forces of supply and demand. In the interim period, it is the responsibility of the purchasing function to minimize the impact of these imbalances.

Economic Conditions

Rapidly changing interest rates and inflation dramatically alter the cost of doing business. The rate of change in the global economy is increasing, due to the transferability of funds that affect the exchange rates and the cost of borrowing. Low interest rates are known to encourage borrowing, promote growth, and spur inflation. Constrictions in the money supply serve to control inflation, but may increase unemployment due to business contraction.

Exchange rate fluctuations and rapid inflation can dramatically affect the purchasing power of the domestic currency. To benefit from these conditions in the global economy, purchasing professionals must constantly monitor the conditions in the global money market. To this end, they may need to enlist the expertise of the finance function of the organization.

Political Factors

Another factor that influences the accuracy of forecasts is the degree of political stability. The United States is considered one of the most stable governments in the world. However, every election brings changes in policies that affect the business community. These include changes in tax rates, employment levels, tariffs, environmental issues, inflation, regulation, and investment rates on the local, national, and international level.

Global Sourcing

Increasing global sourcing opportunities must be evaluated in light of the political risk posed by the host country. Political risk assessments are conducted by the U.S. Department of State, international banks, and multinational organizations. Potential supply shortages and rapid changes in raw material and component prices should be included in the materials forecast. A procurement strategy designed to minimize the risk from political instability in the host country would include a multiple sourcing policy for critical materials. This policy is effective as long as multiple sources are available.

Natural Disasters

A final, uncontrollable factor is the advent of a natural disaster that disrupts the source of supply, such as an earthquake, drought, flood, fire, or explosion. These occurrences cannot be forecasted accurately, but contingency plans can be created. The primary impacts of these emergencies are material shortages and price increases. As is the case with the other threats to material supply, the proactive purchasing professional must be ready to implement contingency plans under such circumstances. The plans should include alternative sources, alternative materials, and stockpiled inventory.

Implications of Forecasting on Procurement

Purchasing effectiveness is dependent on the accuracy and timeliness of forecast information. When presented with a forecast, the purchasing professional should answer the following questions before acting on the forecast:

- Who generated the forecast?
- Who is responsible for the accuracy of the forecast?
- What forecasting methodology was utilized?
- What assumptions are incorporated in the forecast?
- Are these assumptions valid?
- What is the historical accuracy of the forecast?
- What is the length of the forecasting period?
- Have any foreseeable events been omitted from the forecast that would affect its accuracy?
- Does the forecast seem intuitively accurate and reasonable?

It is not wise to accept the forecast on "blind faith" and commit corporate resources without an appropriate amount of scrutiny. This is especially crucial when new products are introduced and no historical data is available, and prior to making long-term contractual commitments to suppliers.

Accurate forecasts foster a synchronized flow of material through the product-delivery transformation process. This results in more effective utilization of corporate resources by fostering long-term planning, enabling the organization to minimize inventory investment, develop stable buyer-supplier relationships, capitalize on volume discounts, and minimize transportation costs. These cost avoidance, profit maximization opportunities are possible due to the reduced uncertainty in the planning environment brought about by forecast accuracy. The Timberland Company is one organization that has reported significant improvements in operating effectiveness due to improved production planning.

> The Timberland Company, the fast growing New Hampshire-based shoe manufacturer, for example, has developed a sophisticated production planning system linked to a sales tracking system that updates demand forecasts. Those systems, along with efforts to reduce lead times in obtaining leather from tanners, have enabled the company to reduce stockout and markdown costs significantly.[20]

Sharing the forecast with members of the supply chain provides additional visibility regarding the long-term volume and timing of requirements. This process serves to minimize uncertainty throughout the supply chain, reducing "firefighting" efforts by suppliers, and promoting the planning of

---

[20]Fisher et al, p. 85.

efficient and effective production scheduling. Therefore, the internal benefits of accurate forecasts can be extended to the entire supply chain. L.L. Bean utilizes its understanding of uncertainty to optimize decisions regarding supply.

> L.L. Bean, the Maine outdoor-sporting goods company, has started to use its understanding of uncertainty to drive its inventory planning decisions. As a direct marketer, Bean finds it easy to capture stockout data. Having discovered that forecasts for its continuing line of "never out" products are much more accurate than those for its new products, Bean estimates demand uncertainty for each category, and then uses those estimates in making product-supply decisions.

Some procurement organizations resist divulging forecast data to their supply base. There is a fear that this information may be leaked to competitors, resulting in a loss of competitive position. Although there is always this possibility, the likelihood of suffering any serious market loss is minimal. In addition, trust is a paramount ingredient in all good relationships, including those with suppliers.

The benefits of early supplier involvement (ESI) far outweigh the potential threats. These benefits will be discussed in detail in Chapter 8. The security provided to the supplier through shared forecast data serves to promote additional investment in product and process improvements. The supplier can evaluate the forecast and determine if the volume requirements justify the investment through amortization of the costs over the life cycle of the product.

In the development of new products, ESI on concurrent engineering teams can result in improvements in product design, process design, cost, quality, and development time. But the primary advantage of ESI in product development is meeting scheduled product launch dates, which are crucial for market penetration and product profitability. This is discussed in greater depth in Chapter 5.

Sharing forecast information on an ongoing basis serves to compress the cycle time associated with the product-delivery process by speeding the flow of information. In the competitive marketplace, time is a significant source of competitive advantage. Organizations are implementing "time-based competitive strategies" to gain market share and enhance profitability. Purchasing can play an important role in time compression by considering time responsiveness in supplier selection decisions, and by working with key suppliers to reduce cycle time.

KEY POINTS

1. Forecasts are a prediction of future events. They act as the catalyst for the entire operating system of the organization. The effectiveness of the operation and the purchasing function is dependent on the accuracy of the forecast.

2. Forecasts may incorporate four basic components: the average level of demand; trends; seasonal influences; and cyclical influences.

3. Economic indicators are used in forecasting because they incorporate activities that change in relation to the economy. There are three groups of economic indicators: leading, coincident, and lagging.

4. The Producer Price Index (PPI), Consumer Price Index (CPI), and the implicit price deflator (GDP deflator) are the indexes most commonly used by the purchasing profession. These indexes adjust data so that comparisons can be drawn over time.

5. Inflation is the measure of sustained price increases in an economy. The inflation rate influences the timing and size of purchases. Rising inflation puts pressure on the buying organization to purchase materials sooner and in larger quantities.

6. External data sources available from the government for use in forecasting include: *The Survey of Current Business, Federal Reserve Bulletin, Statistical Abstract of the United States, Economic Report of the President,* and *Business Conditions Digest.* Associations such as the National Association of Purchasing Management (NAPM), The American Production and Inventory Control Society (APICS), industry associations, and trade associations are also good sources of information.

7. Time series analysis, causal modeling, market research, and qualitative methods represent the four generic forecasting methodologies available to the purchasing professional.

8. It is important to understand the nature of the organization's operating environment prior to selecting the appropriate forecasting tool.

9. Time series analysis is based on using historical data to predict future events.

10. Causal modeling relies on identifying important variables, then identifying cause and effect relationships to predict the future.

11. Market research utilizes data gathering techniques such as surveys, telephone interviews, and live interviews to develop forecasts.

12. Qualitative methods are utilized when no objective data is available. Forecasts are based on the judgment or opinions of experts.

13. Factors that are known to affect forecast accuracy include fluctuations in lead time and labor, changing technology, economic conditions, political instability, supply shortages, and natural disasters.
14. Purchasing effectiveness is dependent on the accuracy and timing of forecast information.

SUGGESTED READINGS

Bretz, Robert S. "Forecasting with Report on Business." *NAPM Insights,* August 1990, pp. 22-25.

Budding, Gonad. "Applying the PPI to Purchasing." *NAPM Insights,* February 1991, p. 6.

Budding, Gonad. "The PPI Regains Influence." *NAPM Insights,* January 1991, p. 26.

Bureau of Labor Statistics Handbook of Methods. U.S. Department of Labor, Bureau of Labor Statistics, Bulletin 2285.

Cuelzo, Carl M., PhD. "Purchasing and Forecasting." *St. Louis Purchaser,* March 1990, pp. 12, 19.

Fisher, Marshall L. et al. "Making Supply Meet Demand in an Uncertain World." *Harvard Business Review,* May 1994, pp. 83-93.

Hoagland, John H., C.P.M., and Barbara E. Taylor, C.P.M. "Purchasing Business Surveys: Uses and Improvements." *Proceedings from 72nd Annual International Purchasing Conference,* May 5, 1987, New York, NY.

"How to Choose the Best Inventory Forecasting Software." *Inventory Reduction Report,* September, 1991, pp. 5-7.

Keen, Howard, Jr. "Use of Weekly and Other Monthly Data as Predictors of the Industrial Production Index." *Business Economics,* January 1988, pp. 44-48.

Klein, Philip A. and Geoffrey H. Moore. "NAPM Business Survey Data: Their Value as Leading Indicators." *Journal of Purchasing and Materials Management,* Winter 1988, pp. 32-40.

Muller, Eugene W., Donald W. Dobler, Harry R. Page, and Eberhard Scheuing. *C.P.M. Study Guide.* Tempe: National Association of Purchasing Management, Sixth Edition, 1992.

Reichard, Robert S. "Index Numbers Boil Down the Trends." *Purchasing,* September 26, 1991, p. 26.

Reichard, Robert S. "Index Users: Proceed with Caution." *Purchasing,* October 24, 1991, pp. 32, 37.

Reichard, Robert S. "Where Buyers Find Price Data." *Purchasing,* March 19, 1992, p. 37.

Reichard, Robert S. "Statisticians Sample Universe." *Purchasing,* December 12, 1991, p. 35.

Torda, Theodore S. "Purchasing Management Index Provides Early Clue on Turning Points." *Business America,* June 24, 1985, pp. 11-13.

Wisner, Joel D. "Forecasting Techniques for Today's Purchaser." *NAPM Insights,* September 1991, pp. 22-23.

Wright, John W., Editor. *The Universal Almanac.* Kansas City: Andrews and McMeel, 1993, pp. 245-54.

CHAPTER 3

STRATEGIC PLANNING PROCESS

This chapter discusses the components of the strategic planning process including formulation, implementation, and control. The chapter begins by defining strategy and discussing the hierarchy of strategies present within an organization.

The concept of environmental "fit" and its relationship to strategic planning is delineated, along with key terminology in the area of strategic management. An overview of the strategic planning process is presented, followed by a discussion of purchasing's role in achieving contemporary, competitive strategies.

DEFINING STRATEGY

The process of strategic planning, while not new to military applications, has only been formerly applied to the business environment since the 1960s.

> Philosopher Sun Tzu, five hundred years before Christ, summed up what's now happening to U.S. industry: "The smartest war strategy allows forces to win without having to fight."[1]

The word *strategy* stems from the Greek word *strategos,* which means "a general" or "to lead." Another derivation, *stratego,* literally defined means:

> To plan the destruction of one's enemies through the effective utilization of resources.[2]

---

[1] Alexander, Earl W. "Operational Overview: Wanted . . . a New Breed of Material Professional." *Purchasing World,* January 1991, p. 28.

[2] Bracker, Jeffrey. "The Historical Development of the Strategic Management Concept." *Academy Management Review,* 1980, Volume 5, Number 2, p. 219.

While this literal definition may seem harsh to the average business professional, the key components can be rephrased and applied to the business environment in the following manner: Strategy is a plan to become the industry leader, involving the effective mobilization and utilization of internal and external resources.

Purchasing's role is to effectively mobilize and utilize the external resources—the supply base.

HIERARCHY OF BUSINESS STRATEGIES

A strategy is a plan to achieve a desired state. The plan represents a blueprint or a "how-to" guide. It defines what needs to be done in order to attain the objective or goal. Consequently, strategy has been defined as encompassing the intended, actual, and unintended actions exhibited by a business entity in the pursuit of an objective.[3]

In the operating realm of an organization, there exists a hierarchy of strategies, including corporate, business, and functional. These different levels in the strategic planning environment correspond to their scope of responsibility and operating domain. The goal is to align the three levels of strategy so they operate in harmony with each other, exhibiting a synergistic effect that benefits the corporation (see Figure 3-1).

Corporate Strategy

Corporate strategy addresses the question, "What businesses do we want to be in?" The corporate strategy encompasses all of the business units owned by the corporation. It can be viewed as similar to portfolio management, with distinct business enterprises operating independently to support one larger framework—the corporation.

During periods of prosperity and good corporate performance, firms have traditionally followed expansionary strategies that include forward and backward vertical integration, related and unrelated diversification, or concentration. These generic strategies are achieved through joint ventures, mergers, market expansion, acquisitions, and internal development (see Figure 3-2). Today, outsourcing and partnering strategies are also being pursued with greater frequency.

---

[3]Mintzberg, Henry and James A. Waters. "Of Strategies, Deliberate and Emergent." *Strategic Management Journal,* 1985, Volume 6, p. 257.

FIGURE 3-1
Hierarchy of Strategies

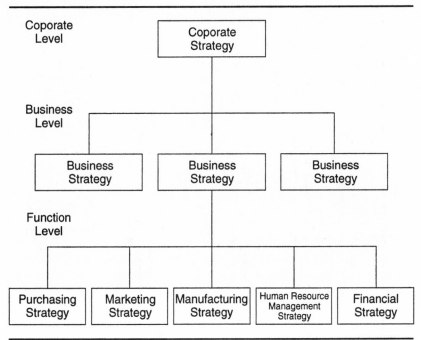

The degree to which a firm is vertically integrated is determined by the breadth of the value-added chain it controls. Forward integration requires a firm to assume more control of the value-added chain in the direction of the final sale to the customer. Backward integration involves gaining control of the supply chain.

Unrelated diversification is a strategic initiative designed to balance the portfolio of businesses under the corporate umbrella. The portfolio would include businesses representing different industries, which ideally do not react to changing economic conditions in the same fashion. This strategy is similar to the one adopted by an individual who elects to invest in mutual funds.

By contract, related diversification seeks to capitalize upon the corporation's core technologies and expertise by selecting strategic business units that are very similar in nature. This strategy opens the corporation to the risks inherent in "putting all your eggs in one basket."

Firms following a concentration strategy grow through market penetration in their core business. This market penetration can be achieved

FIGURE 3-2 Corporate Strategies

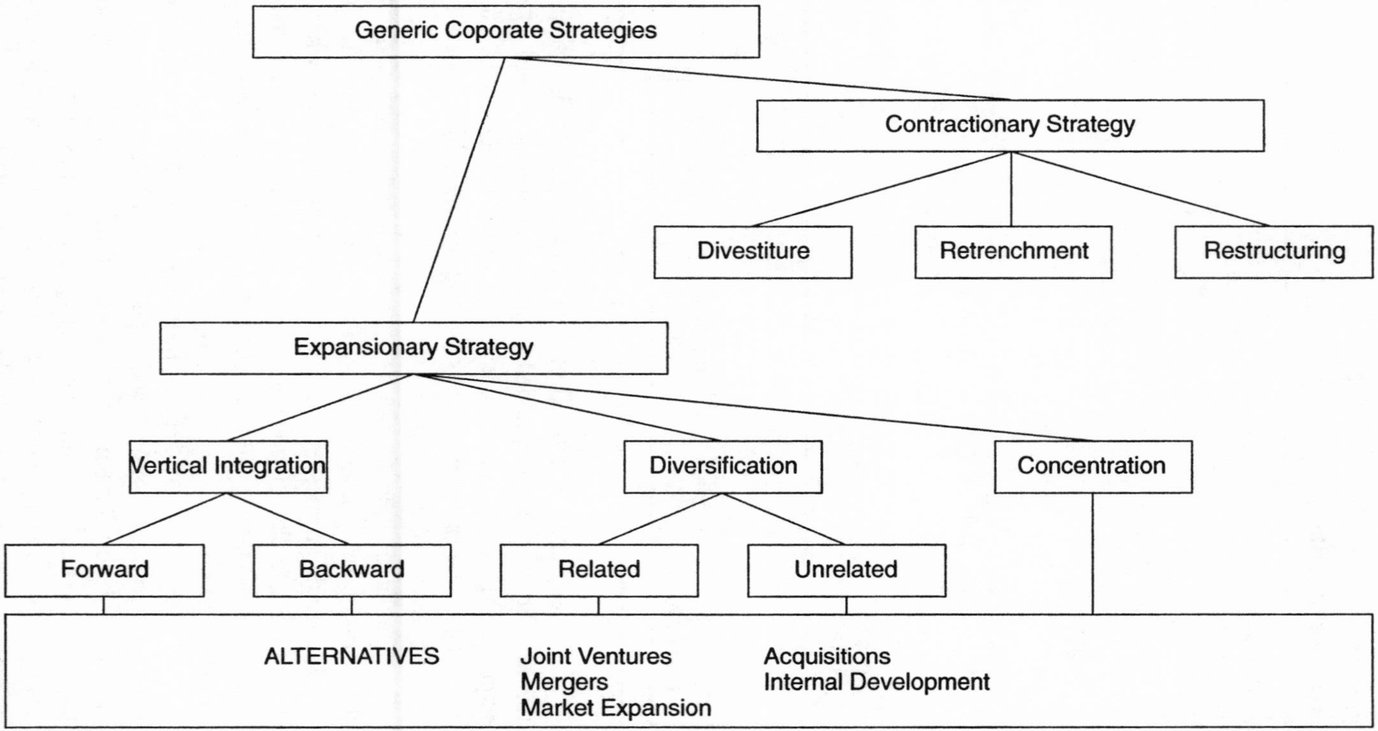

through increasing sales in existing markets, or through growth in previously untapped markets. Timex, the industry leader in watchmaking since 1960, is a primary example of a firm that utilizes a concentration strategy.

> To expand its lineup, Timex in early 1992 bought rivals Guess and Monet Jewelers, gaining a presence in upscale department stores. Early this year it licensed its name for a line of wall clocks and clock radios, due out this summer. A deal with Nautica Apparel Inc. gave Timex its first dressy men's watch. And last month Timex started making watches with Disney movie characters such as Snow White . . . Timex product line now totals 1,500 styles, up from 300 in 1970, ranging in price from $20 to $300 . . . This strategy has brought increases in sales at Timex while the rest of the industry stays flat.[4]

Joint ventures are business arrangements between two or more companies that involve shared ownership. Economic deregulation, an increasing rate of technological obsolescence, increasing global competition, industry and economic maturation, improved communication and computational technology, and industry globalization are the driving forces behind the development of joint venture arrangements.[5]

Examples of joint ventures are common in the automotive industry, including an arrangement between Ford and Mazda and another between Chrysler and Mitsubishi. Joint ventures are taking the form of cross-production, cross-licensing, cross-distribution, cross-development, and cross-national arrangements, in attempts to accelerate the introduction of new products and reduce the risk of development.

Industry-wide research consortiums are an example of cooperative strategies among competitors. They are designed to defray the cost of research and development activities and facilitate the transfer of technology. These research consortiums are usually based on promoting the global competitiveness of a nation like Japan's Ministry of International Trade (MITI).

The 1980s are known as the era of mergers and acquisitions. Acquisitions involve the outright purchase of one firm by another, such as Philip Morris acquiring Kraft Foods, General Motors purchasing EDS, and RJ Reynolds buying Nabisco. Mergers result when two firms join to become a new operating entity. For example, the recent merger consideration

---

[4] Roush, Chris. "At Timex, They're Positively Glowing." *Business Week,* July 12, 1993, p. 141.

[5] Harrigan, Kathryn Rudie. "Strategic Alliances: Their New Role in Global Competition." *Columbia Journal of World Business,* Summer 1987, p. 69.

of Macy's and Federated would have created the largest department store chain in history. Some of these expansionary moves were examples of unrelated diversification, or attempts to diversify the corporate portfolio into businesses different from the firm's primary business interests. Others were examples of related diversification, or efforts to continue expansion within the firm's primary industry.

The primary benefits of an acquisition growth strategy for the buying firm is the ease of entry into new markets and the access to an existing customer base. In addition, the buying firm benefits from the reputation and goodwill that the acquired firm has developed in the market. The predominant risk involves a clash of corporate cultures, as in the case of General Motors and EDS.

Related and unrelated diversification strategies can also be accomplished through internal development by the firm. The development of the Olive Garden restaurant chain by General Mills is a primary example of a firm that followed a related diversification strategy through internal development. Internal development begins with product conception and continues through to delivery to the consumer. It requires financial strength, technical and market expertise, and most importantly, time for the development effort.

> General Mills canvassed 1,000 restaurants for recipes, interviewed 5,000 customers, and tried more than 80 pots of spaghetti sauce before selecting a final service bundle . . . Five years after General Mills opened its prototype's doors, there were 58 Olive Garden restaurants. The firm hopes to eventually operate 500 across the country.[6]

During contractionary business cycles marked by poor corporate performance, firms may elect to follow retrenchment, restructuring, divestiture, or other types of turnaround corporate strategies (see Figure 3-2). Many firms implement such strategies in attempts to improve corporate performance through "downsizing" or "rightsizing". This process results in a redistribution of the labor force, as many employees are asked to take early retirement or are laid off. The United States Post Office, currently under the leadership of its 70th postmaster, Marvin Runyon, is a primary example of an organization that has significantly reduced the number of its employees, eliminating 47,828 workers to date. This represents an effort to reduce the cost of operations by following a retrenchment strategy.

---

[6] Krajewski, Lee, and Larry Ritzman. *Operations Management-Strategy and Analysis.* Reading: Addison Wesley Publishing Company, Inc., 1993, p. 43.

Runyon was out to 'downsize' government before the phrase became popular . . . . Runyon's edict to eliminate 40 percent of the jobs that 'don't touch the mail' proved immediately popular among the agency's principal clients, a politically powerful aggregation consisting these days not of ordinary citizens, but rather of newspapers, direct-mail advertisers, news weeklies, banks and credit card companies. These members of the 'mailing community' (as industry jargon terms them), whose bottom line is strongly affected by postal rates, perceived Runyon as having moved to hold down labor costs and with them the cost of mailing.[7]

Mobil Corporation provides another example of a firm implementing this strategy.

Mobil will take an after-tax charge of $315 million against second-quarter earnings and cut 2,300, or 20 percent, of its chemical workers to boost profits. Chemical profits have fallen due to declining demand.[8]

Restructuring of a corporation involves a change in the structure of the existing portfolio of business units. It may involve combining similar business units, establishing new business units with a unique focus, changing the scope or responsibility of business units, or developing new reporting channels. Eli Lilly, IBM, Apple, and Proctor and Gamble have all announced restructuring plans designed to make them more competitive in their markets.

Eli Lilly said it plans to split five of its medical device and diagnostic businesses to form a new public company, Guidant. If federal regulators approve the plan, the first stock offering is expected during the fourth quarter of this year. The spin-off is part of a restructuring designed to bolster profits hurt by slowing sales of several products, such as the antidepressant drug Prozac.[9]

Another strategy designed to improve corporate performance involves the divestiture of poor performing or unrelated business units, so that the firm can focus its attention on improving the performance of the core businesses. General Motors' recent announcement of its intention to sell National Car Rental is a primary example of this strategy.[10]

---

[7] McAllister, Bill. "Can Marvin Runyon Deliver?" *The Washington Post Magazine.* July 10, 1994, p. 28.

[8] Joy, Pattie and Shakira Hightower. "Mobil Cuts Jobs." *USA Today,* June 29, 1994, p. 1B.

[9] Joy, *USA Today,* June 29, 1994, p. 1B.

[10] Jones, Del. "GM to Sell National Car Rental." *USA Today,* June 23, 1994, p. 1B.

Business Strategy

While corporate strategy focuses on determining what businesses the firm should be in, business level strategy is designed to answer the questions, "How do we compete in this industry?" This involves an analysis of the competition, and an interpretation of the role each competitor fills in the industry. A widely accepted strategic framework developed by Michael Porter (1980) identifies three generic strategies for competing in an industry: cost-leadership, differentiation, and focus.

The cost-leadership strategy is accomplished by being the lowest-cost producer in the industry. Some examples of firms from various industries that have achieved this distinction include Hyundai Motors (automotive manufacturing), White Castle (fast food), Earl Scheib (automobile painting), and Red Roof Inns (motel industry). Firms utilizing this strategy obtain advantages through an ability to compete in the market on the basis of price, or by enjoying higher profit margins than competitors at a market-determined price.

Firms capable of creating a "perceived" difference in the customer's mind on some product attribute may utilize a differentiation strategy. Research has shown that a differentiated producer can charge up to ten percent more for its product than the competition.[11] It is important to remember that the difference in the product does not have to be tangible or real, as long as there is a *perceived* distinction in the consumer's mind.

Some typical characteristics that firms utilize to differentiate their products include quality, performance, status, and sex appeal. Firms noted for competing on differentiated quality include Mercedes Benz and the Ritz Carlton. Porsche and Nikon are distinguished by their high performance product designs. Status is conveyed through the cost of ownership of a Rolls Royce or a Rolex watch, while heightened sex appeal is the selling feature for Guess jeans and Calvin Klein perfume.

The previous two generic strategies, cost-leadership and differentiation, are intended for broad-scale applications in terms of target market segmentation. They are normally utilized on a national or international basis. The remaining generic strategy, focus, is reserved for use with a target market that is restricted in size. The restriction may be based on any market segmentation criteria, including income levels, geographic distribution, gender, age, education, and others. A familiar marketing term utilized to

---

[11] Phillips, Stephen. "King Customer." *Business Week,* March 12, 1990, p. 90.

describe this strategy is "niche." Micro-breweries, local moving companies, and many service organizations follow a focus strategy. Firms compete in a niche by utilizing the cost-leadership or differentiation strategy, but the distinguishing characteristic is the scope and scale of operations, which is usually local or regional in nature.

Functional Strategy

Functional level strategies include purchasing, marketing, customer service, finance, accounting, human resource management, operations, and research and development. The inherent diversity of functional level strategies prevents the development of a comprehensive definition. It is sufficient to understand that effective functional level strategies are plans that support the attainment of business and corporate objectives.

Within the purchasing function, the strategic focus converges on integrating the various activities into the total corporate scheme, and designing strategic programs closely aligned with current and anticipated environmental changes.[12]

Positive corporate performance over the long-term requires the proper alignment of the business enterprise, including strategies, culture, and systems.[13] Corporate, business, and functional level strategies must be "nested" in an interlocking fashion to foster the desired state of synergy.

Nested strategies recognize the direction, parameters, and objectives established by the preceding level in the hierarchy of business strategies, from corporate to functional (see Figure 3-1). The higher-level strategy serves to set the scope of operations, objectives, goals, and resource availability for subsequent strategic planning activities on the lower levels in a trickle-down process. The strategies must be supported by a culture that demonstrates the values of the corporation, and by informational, structural, and rewards systems that act as facilitating agents.

Concept of Strategic Fit

Corporate performance has also been linked to the concept of "fit." Just as performance is related to internal alignment, it is also related to external alignment, or the relationship between the business entity and the external operating environment. Firms that demonstrate the appropriate "fit" thrive in

---

[12] Spekman, Robert E. "A Strategic Approach to Procurement Planning." *Journal of Purchasing and Materials Management,* Winter 1981, p. 5.

[13] Covey, Stephen R. *Principle-Centered Leadership.* Fireside New York: Simon & Schuster, 1991.

their environment. Firms out of alignment, or firms that fail to adapt to a changing environment, will not prosper, and may not survive under competitive market pressures. This process of weeding out, or "survival of the fittest," is drawn from the theory of evolution, and is referred to as "industrial Darwinism." Successful firms take on the characteristics of chameleons or amoebas, in the sense that they are flexible and capable of adapting to the changing demands of the external operating environment.

Peak performance therefore involves the strategic alignment of the entire business system—both internal and external. Contemporary business literature refers to these linkages as components of the value-added chain. It recognizes that every business system is composed of two markets—the supply side and the market side. The strategic planning process therefore begins with an analysis of both the external environment, through an in-depth industry analysis, and the internal or company environment, through an accurate assessment of the firm's capabilities and limitations.

STRATEGIC PLANNING FRAMEWORK

The Nature of Markets

The strategic planning process requires an understanding of the composition of markets and the nature of competition present in the industry. The United States economy is composed of five basic industries, including: basic manufacturing (automobiles, steel, machinery, textiles, appliances); energy (natural resources, technology, facilities); high technology (semiconductors, computers, industrial and home electronics); agriculture (growers, packers, distributors of "foodstuffs"); and services (health, financial, personal, communications, information, entertainment). These industries are known as subeconomies, because they each display unique market dynamics governed by the interplay of supply and demand.

In economies where the price of goods or services is not set by the political administration or a socialist government, or regulated with price controls, price will be determined by the levels of supply and demand present in the market. Supply and demand within an industry are characterized by their level of "elasticity." Elasticity is a measure of the responsiveness of the supply and demand functions to changes in price.

A product that is highly elastic will demonstrate pronounced changes in the level of demand with changes in the price of the product.

A reduction in price would cause a significant increase in the level of demand. Items that fall into this category are often referred to as luxury goods or services. Inelastic or "unitary elastic" products demonstrate relatively minor changes in demand relative to fluctuations in price. Basic necessities such as clothing, shelter, and food, which are essential for human survival, are considered inelastic. Consumers may substitute products, but the basic need cannot be postponed or eliminated.

Figure 3-3 presents a graphic depiction of the demand (D_0) and supply (S_0) curves with the associated market equilibrium price determined by their intersection (P_0). The demand curve represents the summation of all the independent demand for the product at various price levels, and is known as the *market demand curve*. The supply curve represents the aggregate production levels of the industry at the corresponding price levels.

While basic needs remain relatively constant over time, shifts in demand can result from changing consumer preferences or levels of disposable income. Figure 3-4 demonstrates the dynamic nature of the supply and demand markets. As demand (D_1) increases, indicated by a shift in the demand curve out and to the right, while supply (S_0) remains constant, prices (P_1) will rise. Over time, the supply curve, attracted by the higher market price, will respond to this change in demand by making more product available. This in turn will establish a new equilibrium price.

Consumer consumption levels are influenced by the "Law of Diminishing Marginal Utility." The utility to the consumer is the added

FIGURE 3-3
Economic Forces of Supply and Demand

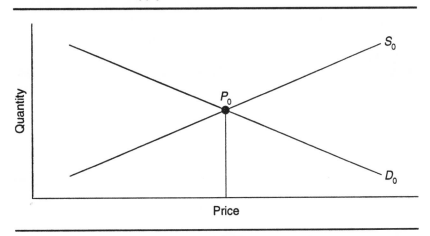

FIGURE 3-4
Shifts in Demand

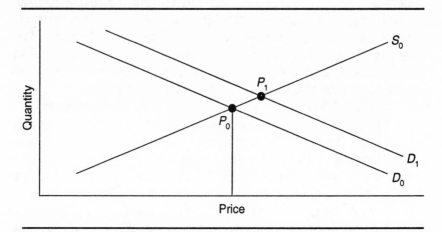

value of the last unit purchased. Marginal utility is the incremental value added by each successive unit purchased. For instance, if you are very hungry, the utility of the first mouthful of food is higher than the next, and much higher than the last. The marginal utility of each mouthful of food declines as the level of hunger diminishes.

Production levels are influenced by the "Law of Diminishing Marginal Returns." Marginal returns are the incremental profit added for each unit produced. In situations where supply outpaces demand, the equilibrium price is driven down. The per-unit profit therefore declines, resulting in the phenomenon of diminishing marginal returns.

Producers are encouraged to supply quantities that meet or exceed their break-even point, which is determined by the intersection of their total revenue (TR) and total cost (TC) production functions (see Figure 3-5). Total revenue is calculated by multiplying the quantity sold by the associated price. Total cost is the aggregation of the fixed and variable components of cost. Fixed costs are those that do not fluctuate with changes in production quantities, such as salaries, rent, insurance, and property taxes. Variable costs are those that change in relation to the production volume, including direct labor, direct material, and energy.

Below this break-even quantity the total cost exceeds the total revenue, so the producer incurs a loss. Profits are made at quantities produced and sold above this equilibrium point, as total revenue exceeds total cost. Producers cannot stay in business indefinitely at levels below their break-

FIGURE 3-5
Break-Even Analysis

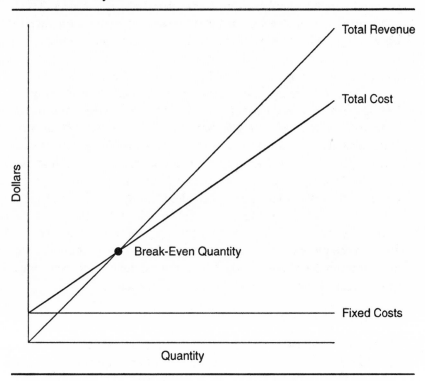

even point. They will, however, continue to produce for a period of time at quantities below their break-even point, because doing so helps them recover at least their variable costs.

The behavior of markets is also influenced by the level of competition between suppliers in the industry. Markets are often described as either a monopoly or an oligopoly, or by referring to perfect or imperfect competition. Each of these categories is distinguished by the number of suppliers present in an industry.

In a monopoly, there is only one supplier for the entire market, a situation that puts the supplier in a very powerful position to control price. Utilities in many locations operate as monopolies that are regulated by government. American Telephone & Telegraph (AT&T) was the only supplier of telephone equipment and services in many regions of the U.S. until the deregulation of the industry.

The other extreme is the case of perfect competition, in which a potentially infinite number of suppliers serve the market. Under these circumstances, no supplier has enough influence in the market to affect the price of the product or service. Industries operating under such conditions are termed fragmented industries, and they are represented by services including hair styling, child care, dry cleaning, and housecleaning.

The term oligopoly is used to describe markets that are supported by a limited number of suppliers. This condition can arise when scarce commodities are under the control of a few suppliers, as in the case of diamond production. Oligopolies may also form when the cost of entering and competing in the industry is very high. Examples include the automobile and aerospace industries.

While all of the preceding market conditions exist to varying degrees, the most dominant market condition is the state of imperfect competition. In this situation, many suppliers provide products or services that have real or perceived differences according to the consumer. The differences in the product offering serve as a source of distinction and create asymmetry in the market. Producers use the differences between competitors to create a competitive advantage in the market.

International Markets

Understanding the underlying dynamics of markets has become more important with the removal of trade barriers through international agreements and trading partner arrangements, including the North American Free Trade Agreement (NAFTA), the General Agreement on Tariffs and Trade (GATT), and the European Economic Community (EEC). Access to new markets for sourcing, as well as selling, has been further enhanced by the end of the Cold War. These changing market conditions present many new opportunities for the business community and the purchasing professional.

According to economic theory postulated by Adam Smith in *The Wealth of Nations* (1979), reduction in trade barriers will result in a redistribution of world resources as the "invisible hand" operates to maximize the efficiency of market economies. This phenomenon will occur as nations seek to capitalize on their *absolute* advantage, which is derived from the ability to produce a unit of output with fewer resources than any trading partner.

In the absence of an absolute advantage, nations will strive to maximize their *comparative* advantage, which is the efficiency advantage they enjoy over other nations producing similar products or services. Regions or nations will therefore specialize in producing those goods or services that yield the highest degree of competitive advantage in the world market.

Artificial comparative advantages can be created by government subsidies or by trade barriers that nullify price differentials in the market. The government of France has been subsidizing Airbus, so that it can effectively compete in the world market with the likes of Boeing and McDonnell Douglas. These artificially created comparative advantages have also led to a practice known as "dumping," in which a producer will actually sell a product or service at below domestic market prices. The United States computer industry (chip producers) and the automotive industry (minivans) are notable victims of dumping as practiced by foreign manufacturers.

The flow of goods and services between nations in the world market is not only governed by trade barriers, but also by the exchange rate. The exchange rate is the value or purchasing power of the domestic currency in a foreign market/currency. At one time, all currency was converted into its value in gold for comparative purposes and to determine the exchange rate. This was known as the gold standard. Today, exchange rates are established by central banks and are allowed to "float" within a given set of parameters, also allowing them to react to changing market and economic activity. A "strong" currency enhances a nation's purchasing power in the international market, but also makes it difficult to sell domestic products, because the selling price is high relative to the "weak" foreign currency.

Currency exchange rates affect the flow of goods and services in and out of a country, as measured by the level of import and export activity known as the "balance of trade." An economy is considered to have a favorable balance of trade when it exports more than it imports. The "balance of payments" figure combines the balance of trade with other accounts, such as tourist spending and foreign investment. The United States economy has not reflected a favorable balance of trade or payments since 1950.

STRATEGIC PLANNING PROCESS

External Analysis

As previously stated, the strategic planning process requires an analysis of both the external and internal operating environments. Examining the external environment requires a full investigation on the industry level. Michael Porter (1980)[14] provides a framework for understanding the

---

[14]Porter, Michael. *Competitive Strategy.* New York: Academic Press, 1980.

dynamics present in any industry. His framework is known as the "Five Force Model." Figure 3-6 depicts the five forces in Porter's conceptual model, including suppliers, new entrants (new competitors in the industry), buyers, substitutes (potential substitute products), and industry competitors. The table shows the major factors that are known to influence the power of the associated force in the industry. The theory holds that the lower the power of each force in the industry, the higher the industry profit potential.

For purchasing professionals, understanding the relative power of each force in their own industry is critical for successful negotiations and establishing effective intercompany relationships. Specific attention should be focused on the balance of power between the buyer and the supplier, and the determinants of their power (see Figure 3-7). Actions can then be taken to change the balance of power so the relationship provides positive results for the purchasing organization. Some strategies that can serve this purpose include changing the percentage of business done with a supplier, increasing the number of alternative sources and substitutes, or backward integration, in which the company may become its own supplier.

Internal Analysis

The internal analysis begins with the organizational mission statement. The mission statement reflects the organization's purpose, operating philosophy, and

FIGURE 3-6
Porter's Five Force Model

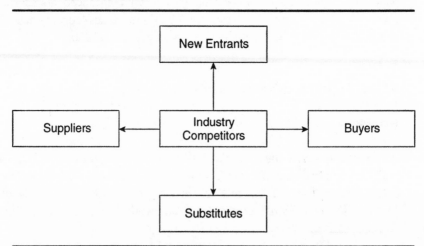

Source: Michael E. Porter, *Competitive Strategy,* New York: The Free Press, 1985, p. 5.

FIGURE 3-7
Determinants of Supplier Power

Force	*Determinants of Supplier Power*
Suppliers	Input Differentiation
	Costs of Switching Suppliers
	Presence of Substitute Inputs
	Supplier Concentration
	Importance of Volume (to supplier)
	Cost Relative to Total Purchases in Industry
	Impact of Inputs on Cost of Differentiation
	Threat of Forward Integration
Substitutes	Relative Price Performance
	Switching Costs
	Buyer Propensity to Substitute
Industry Competitors	Industry Growth
	Fixed Costs/Value Added
	Intermittent Overcapacity
	Product Differences
	Brand Identity
	Switching Costs
	Concentration and Balance
	Informational Complexity
	Diversity of Competitors
	Corporate Stakes
	Exit Barriers
Barriers to New Entrants	Economies of Scale
	Proprietary Product Differences
	Brand Identity
	Switching Costs
	Capital Requirements
	Access to Distribution
	Absolute Cost Advantages
	Government Policy
	Expected Retaliation
Buyers	Bargaining Leverage
	Concentrations (buyer versus firm)
	Buyer Volume
	Buyer Switching Costs
	Buyer Information
	Ability to Backward Integrate
	Substitute Products
	Price Sensitivity
	Price/Total Purchases

Source: Michael E. Porter, *Competitive Strategy,* New York: The Free Press, 1985, p. 6.

values. It offers the broadest sense of direction to guide decision making in the organization. The Declaration of Independence represents the mission statement for the entity known as the United States. Mission statements are relevant to the corporate, business, functional, and personal levels of an organization.

From the mission statement, organizational objectives are developed. The objectives are the desired state that the organization wishes to attain. Specific goals are then established, based on these overriding objectives. The goals differ from the objectives in that they state tangible, measurable results the organization intends to achieve. For example, one objective for a corporation may be to increase market share. The goal corresponding to this objective might be stated as: "Increase market share in the portable printers segment by three percent."

In order to develop the appropriate strategies and tactics necessary to achieve the goals and objectives and support the mission of the organization, a SWOT analysis is usually conducted. SWOT stands for strengths, weaknesses, opportunities, and threats. The strengths and weaknesses are an assessment of the internal organizational capabilities. The opportunities and threats are those conditions present in the current and future operating environment of the firm.

At this juncture, the concept of strategic "fit" comes into play. The fit involves maximizing the alignment of the internal strengths of the firm and the external opportunities present in the industry, and minimizing the firm's risk from external threats and internal weaknesses. Strategies represent the specific plans designed to accomplish and achieve the corporate objectives.

While strategies detail the "how-to" aspects of accomplishing a desired objective, tactics make the strategies operational by specifying actions that lead to implementation. Tactics outline which resources will be used to implement the strategy and the time frame for accomplishing the goal.

The implementation plan for any strategy must include methods of measurement, evaluation, monitoring, and control. Prior to implementing the strategy, appropriate measurements of success must be established and linked to specific goals. The organization needs a way of collecting the data, evaluating the results, monitoring progress toward the goal, and controlling the outcomes of the plan.

General Economic Issues

Strategic planning must take into account the broader macroeconomic environment in which the firm operates. Some of the issues affecting the organization's operating environment involve national and international regulations. The following section will highlight some of the issues that must be considered in developing strategic plans.

Transportation Trends

The transportation industry in the United States was relatively stable from the late 1940s to the 1970s. Rates were uniform, and changed in small, predictable amounts each year. Since the mid-1970s, rising fuel and capital costs and supply shortages have disrupted the stable carrier situation. Regulatory reforms of the late 1970s, commonly referred to as deregulation, have also brought tremendous changes to the industry.

Some of the changes caused by regulatory reform are increasing the importance of inbound transportation decisions. Some industry trends include: companies reducing their carrier base; deep discounting of rates awarded by new entrants; a large number of entries and exits from the industry; a need for closer relationships, including partnerships between buyers and carriers as well as closer relationships between purchasing and traffic/transportation departments; carriers offering more innovative rates and services; and companies seeking to outsource the transportation function.

It is expected that in the future there will be a continued concentration in all forms of transportation; increased use of intermodal transportation; the emergence of mega transportation companies that will provide a broad range of services; and increased use of third-party transportation companies.

Money Supply

The money supply refers to the total money available to a nation for spending. It consists of coin, currency, and demand (checking) accounts held by private businesses and individuals. The three commonly used measures of the money supply are M-1, M-2, and M-3. M-1 consists of currency plus demand deposits. M-2 is M-1 plus time deposit balances (savings accounts). M-3 consists of M-2 plus time deposits (such as certificates of deposit) held at commercial banks and savings banks.

Interest Rates

The interest rate, or the price of borrowing money, fluctuates according to a variety of factors. These include the lender's willingness to postpone use of the money, the lender's willingness to assume risk, the possibility that economic changes will reduce the purchasing power of funds by the time a loan is repaid, and the administrative cost of processing a loan. The demand level for a particular type of loan and the supply of funds available for lending also affect interest rates.

Inflation

Inflation is the result of a general rise in prices and wages in response to changes in consumption, savings, and investment levels. Inflation represents a loss of purchasing power due to price increases for needed items. Political and economic developments influence price and wage levels. Demand-pull inflation means that prices and wages are pulled up by the demand for limited supplies of material and labor.

Inflation can be mild, chronic, or hyper. Hyperinflation occurs after war or revolution, when prices may multiply by hundreds of times. Chronic inflation occurs when high inflation rates of 30 percent, 40 percent, or more are present year after year. Sudden rises in inflation cause problems in the economy, because they affect interest rates and organizations and individuals on fixed incomes.

Employment Levels

The employment level is usually expressed as the percentage of people unemployed at any given time. Structural unemployment occurs because the size and composition of the labor force is continually changing as new people join the workforce and others retire or change careers. About four percent of the labor force is structurally unemployed at any given time. Full employment, then, really means 96 percent employment.

The Employment Act of 1946 mandates that the federal government do all it can to ensure full employment. Two means that the government can utilize to affect employment are fiscal policy and monetary policy. The government shapes taxation and public spending through its fiscal policy, and can thereby dampen swings in the business cycle and maintain high employment. Monetary policy affects interest rates and shapes consumption and savings. It can also influence the economy.

Government Budgeting

The increase in federal spending over the last 35 years has far exceeded the growth in the U.S. tax base. Government program growth has been financed in part by legislated and automatic tax increases. Since tax increases alone have been insufficient to finance the large growth in spending, the U.S. has experienced unprecedented peacetime budget deficits in recent years.

The growing budget deficit has led to increased uncertainty about economic policy, which is reflected in interest rates and thus investments. The government's need for borrowed money affects interest rates and competes with the needs of basic industries for capital to finance new plants and equipment. The federal deficit thus reduces the economy's growth potential and its competitiveness at home and abroad.

Issues in International Procurement

Free Trade Versus Protectionism
The term "free trade" refers to a situation in which there are no tariffs, no quotas, and no embargoes. "Fair trade" implies some degree of protectionism, such as selective and graduated tariffs and negotiated quotas on selected products. Protectionist measures are also instituted for government buying.

Countertrade
Countertrade is essentially any transaction in which all or part of the payment is made in goods instead of money. The need for countertrade is driven by the balance of payments problems of a country, and by weak demand for the country's products. A likely candidate for countertrade is a country with a shortage of foreign exchange or a shortage of credit to finance trade flows. Such a country will be trying to expand its exports or develop markets for its new products. In their book, *Creative Countertrade*, Elderkin and Norquist refer to five basic forms of countertrade: barter, buyback, compensation, counterpurchase, and switch.[15]

Barter, the simplest form of countertrade, occurs when goods of equal value are exchanged and no money is involved. In buyback arrangements, the selling firm provides equipment or an entire plant, and agrees to buy back a certain part of the production. Many less developed countries insist on buy back arrangements, because they ensure access to Western technology and stable markets. The amount of compensation is specified as a percentage of the value of goods being traded, to the value of the product being sold, referred to as an offset arrangement. Compensation is actually offset, but specifies the percent of the value of goods being countertraded to the value of the product being sold. Counterpurchase involves more cash in the transaction, smaller volumes of goods flowing to the multinational corporation over a shorter period of time, and goods unrelated to the original deal. A switch transaction uses at least one third party outside the host country to facilitate the trade. The countertraded goods or the multinational's goods are sent through a third country, for purchase in hard currency or for distribution. While countertrade agreements may be complex, they do offer an opportunity to develop lower-cost sources of supply in the world marketplace. In some cases, they may provide the only means of market entry for the firm.

---

[15] Elderkin, Kenton W., and Warren E. Norquist, *Creative Countertrade*, Cambridge: Ballinger Publishing Company, 1987, p. 152.

Resources For Identifying International Suppliers.
Identifying potential sources of supply in the international market can be
facilitated by utilizing one of the following:

Brokers. Brokers are paid commissions by sellers to locate buyers,
and are paid by buyers to locate sources of supply. Brokers, however, are
not involved in the shipment or clearance of an order through customs.
Nor do they assume any of the seller's fiscal responsibility.

Consuls, Embassies, Missions. Almost all countries in the world maintain
an embassy in Washington, D.C. Major industrial nations may also maintain
trade consulates in the U.S. These places will provide names of suppliers and
background information if their role is to promote exports from their country.

Banks. Most major banks have a foreign trade department, which
will provide information that is helpful in locating potential sources, as
well as information on currency, payment, documentation, and governmental approval procedures.

Export Trade Companies/Associations. These organizations handle a
wide spectrum of products from one or a limited number of countries. They
are used extensively by Japanese firms to move products into North America.

Chambers of Commerce. The U.S. Department of Commerce has
current lists of names and addresses of foreign suppliers, organized by
general type of products. The International Chamber of Commerce has
contacts around the world through its country branches, and it can supply
leads on possible sources.

A glossary of terms relevant to international procurement is presented
on page 63.

KEY POINTS

1. A strategy is a plan to become the industry leader. It involves the
   effective mobilization and utilization of internal and external resources.
2. A hierarchy of strategic levels exists in any organization, including the
   corporate, business, and functional levels.
3. Corporate strategy addresses the question: "What businesses do we
   want to be in?"
4. Generic corporate expansionary strategies include forward and backward vertical integration, related and unrelated diversification, and
   concentration. Contractionary corporate strategies include retrenchment, restructuring, and divestiture.

5. Business strategy is designed to answer the question: "How do we compete in the industry?"

6. The three generic business strategies identified by Michael Porter include cost-leadership, differentiation, and focus.

7. Effective functional strategies support the attainment of business and corporate objectives.

8. Corporate performance is enhanced by strategies that support each other in a "nested" fashion, creating a synergistic effect.

9. Strategic "fit" recognizes the importance of the relationship between the organization and its operating environment. Proper alignment with the environment is essential to long-term success.

10. Markets can be analyzed by understanding the dynamic nature of supply and demand within an industry and the level and type of competition that is present.

11. The strategic planning process begins with a thorough analysis of the external environment, at the industry level, and the internal environment, at the corporate level.

12. General economic issues that impact strategic planning include transportation trends, the money supply, interest rates, inflation, employment levels, and the level of government spending.

13. Countertrade is a transaction commonly involved in international procurement. It can take the form of barter, buyback, compensation, counterpurchase, and switch.

GLOSSARY OF INTERNATIONAL PROCUREMENT TERMS

Ad valorem duty rate.   Ad valorem is the most common U.S. Customs duty, or tariff. It is based on the value of the goods involved (usually a fixed percentage). For example, if the ad valorem tax on a radio is 15 percent and the radio costs $60, the buyer must pay a duty of $9 per radio, making the total cost $69.

C&F (CFR), or "Cost and Freight."   This is a shipping payment term indicating that the seller is obligated to pay the costs and freight necessary to transport the goods to the named destination. The risk of loss or damage is transferred from the seller to the buyer when the goods pass the ship's rail in the port of shipment.

CIF, or "Cost, Insurance and Freight."   This term is essentially the same as C&F, with the addition that the supplier has the responsibility to procure maritime insurance against the risk of loss or damage during carriage.

Charter rate. A charter is a fee charged for leasing a ship (usually a tanker) for a voyage, or for leasing cargo space aboard the vessel. Charter rates tend to fluctuate widely.

Compound duty rate. A compound rate is a combination of both an ad valorem and a specific rate.

Customs broker. A customs broker is a licensed professional who, for a fee, will act on a buyer's behalf in the dealing with the U.S. Customs Service to bring imports into the country.

Specific duty rate. This is a customs duty rate that is a specified amount per unit of weight or other unit of measurement.

SUGGESTED READINGS

Adamson, Joel. "Corporate Long-Range Planning Must Include Procurement." *Journal of Purchasing and Materials Management,* Spring 1980, pp. 25-32.

Alexander, Earl W. "Operational Overview: Wanted . . . a New Breed of Material Professional." *Purchasing World,* January 1991, pp. 28-29.

Billington, Corey. "Strategic Supply Chain Management." *OR/MS Today,* April 1994, pp. 20-27.

Bracker, Jeffrey. "The Historical Development of the Strategic Management Concept." *Academy Management Review,* 1980, Volume 5, Number 2, pp. 219-24.

Burt, David N. "Managing Suppliers Up to Speed." *Harvard Business Review,* July/August 1989, V. 67, I. 4, pp. 127-35.

Caddick, J.R., and B.G. Dale. "The Determination of Purchasing Objectives and Strategies: Some Key Influences." *International Journal of Physical Distribution & Materials Management,* Volume 17, Number 3, pp. 5-16.

Chandler, A. *Strategy and Structure.* Cambridge: Harvard University Press, 1962.

Cooper, Martha C., and Kevin Humphreys. "The 'How' of Supply Chain Management." *NAPM Insights,* March 1994, pp. 30-32.

Covey, Stephen R. *Principle-Centered Leadership.* New York/Fireside: Simon & Schuster, 1991.

Ellram, Lisa M. "The 'What' of Supply Chain Management." *NAPM Insights,* March 1994, pp. 26-27.

Farmer, David H. "Developing Purchasing Strategies." *Journal of Purchasing and Materials Management,* Fall 1978, pp. 6-11.

Farmer, David H. "Seeking Strategic Involvement." *Journal of Purchasing and Materials Management,* Fall 1981, pp. 20-24.

Fox, Harold W., and David R. Rink. "Coordination of Purchasing with Sales Trends." *Journal of Purchasing and Materials Management,* Winter 1977, pp. 10-18.

Hahn, Chan K., Kyoo H. Kim, and Jong S. Kim. "Costs of Competition: Implications for Purchasing Strategy." *Journal of Purchasing and Materials Management,* Fall 1986, pp. 2-7.

Harrigan, Kathryn Rudie. "Strategic Alliances: Their New Role in Global Competition." *Columbia Journal of World Business,* Summer 1987, pp. 67-69.

Jones, Del. "GM to Sell National Car Rental." *USA Today,* June 23, 1994, p. 1B.

Joy, Pattie. "Eli Lilly Spinoff." *USA Today,* June 21, 1994, p. 1B.

Joy, Pattie and Shakira Hightower. "Mobil Cuts Jobs." *USA Today,* June 29, 1994, p. 1B.

Kiser, G.E., and David Rink. "Use of the Product Life Cycle Concept in Development of Purchasing Strategies." *Journal of Purchasing and Materials Management,* Winter 1976, pp. 19-24.

Krajewski, Lee, and Larry Ritzman. *Operations Management-Strategy and Analysis.* Reading: Addison Wesley Publishing Company, Inc., 1993.

Kraljic, Peter. "Purchasing Must Become Supply Management." *Harvard Business Review,* September-October 1983, pp. 109-17.

McAllister, Bill. "Can Marvin Runyon Deliver?" *The Washington Post Magazine,* July 10, 1994, pp. 16-35.

Miller, Jill."The Customer, the Purchaser, and the Supplier: Is It a Three-Way Street." *NAPM Insights,* May 1994, pp. 30-32.

Mintzberg, Henry and James A. Waters. "Of Strategies, Deliberate and Emergent." *Strategic Management Journal,* 1985, Volume 6, pp. 257-72.

Morgan, James P. "When Sourcing Begins to Drive Corporate Strategy." *Purchasing,* January 17, 1991, pp. 122-29.

Murphree, Julie. "Learning to Learn Together." *NAPM Insights,* October 1992, pp. 14-15.

Murphree, Julie. "Back to the Starting Point with the Final Customer." *NAPM Insights,* May 1994, pp. 28-29.

Murphree, Julie. "How Far Have Teams Come?" *NAPM Insights,* July 1994, pp. 26-27.

Napoleon, Landon J. "Increasing the Value." *NAPM Insights,* May 1994, pp. 37-40.

Newman, Richard G. "Single Sourcing: Short-Term Savings Versus Long-Term Problems." *Journal of Purchasing and Materials Management,* Summer 1989, pp. 20-25.

Pearce II, John A. "The Company Mission as a Strategic Tool." *Sloan Management Review,* Spring 1982, pp. 15-24.

Phillips, Stephen. "King Customer." *Business Week,* March 12, 1990, pp. 88-94.

Porter, Michael. *Competitive Strategy.* New York: Academic Press, 1980.

Raia, Ernest. "1989 Medal of Professional Excellence: Supply Line Management is Taking Purchasing Beyond the Normal Boundaries of Supplier Relations." *Purchasing,* September 28, 1989, pp. 51-66.

Reck, Robert F., and Brian G. Long. "Purchasing a Competitive Weapon." *Journal of Purchasing and Materials Management,* Fall 1988, pp. 2-8.

Roush, Chris. "At Timex, They're Positively Glowing." *Business Week,* July 12, 1993, p. 141.

Smith, Adam. *The Wealth of Nations,* Books I-III, Middlesex, England: Penguin Books, Ltd., 1979.

Spekman, Robert E. "A Strategic Approach to Procurement Planning." *Journal of Purchasing and Materials Management,* Winter 1981, pp. 2-7.

Treleven, Mark. "Single Sourcing: A Management Tool for the Quality Supplier." *Journal of Purchasing and Materials Management,* Spring 1987, pp. 19-24.

CHAPTER 4

MANAGING INTERNAL RELATIONSHIPS

Purchasing is one of many functional areas that make up any organization. In order for the organization to achieve its overall objectives, the purchasing function must work in harmony with other functions, through good two-way communication, common goals, and mutual respect. Organizations today are increasingly attempting to reduce the amount of functional loyalty and functional territorialism by using work groups that cut across functional boundaries.

This chapter begins with a discussion of basic communication and the perceptual issues that affect the purchasing function's relationship with other departments. The next section discusses specifically how the purchasing function is related to other functions within the organization. The emphasis is on the benefits of mutual cooperation. The chapter closes with a look at the way organizational structures have evolved.

INTERFUNCTIONAL RELATIONSHIPS

Virtually every department within an organization relies on the purchasing function for some type of information or support. Purchasing's role ranges from a support role to a strategic function. To the extent that purchasing provides value to other functional areas, it will be included in important decisions and become involved early in areas that affect purchasing. Being well-informed allows the purchasing function to better anticipate and support the needs of other functional areas. This support in turn leads to greater recognition and participation.

Because the purchasing function is a major supplier and user of internal information, excellent interfunctional relationships are critical to the success of the purchasing function. Many organizations are aware of the importance of maintaining good internal relationships. For example, purchasing employees at Corning, Inc., and the Army and Air Force Exchange Service have begun to call the internal functions that rely upon them their "customers" or "internal customers." AT&T purchasing uses the term "internal clients." Treating internal functions as customers creates a new mind-set. Purchasing views other functions not as adversaries, but as parties to be served and supported. Such a change can have a major impact on improving internal relationships and increasing purchasing's involvement in critical areas. Once the purchasing function has proven itself as a vital, contributing function, the doors open for further participation. The role and the status of the purchasing function is elevated within the organization.

Communication Vehicles

The purchasing function communicates with other areas in many ways. Key factors to consider are the communication mechanisms, who is receiving the information, and what their needs are. The way the purchasing function is perceived within the organization is also important, as is establishing mutual trust between purchasing and other departments.

Communication Mechanisms
Purchasing communicates with other areas verbally and in writing. Purchasing is the recipient and sender of many reports. Purchasing may receive and communicate via stores and inventory status reports, receiving documents, production needs, and purchase orders. These reports may be rather routine and taken for granted.

Other, less routine information flows include engineering and design change orders, specifications, and specification changes. These may require careful scrutiny and a deep understanding in order to assure that needs are met, and that opportunities for reducing costs and standardizing have been identified by the purchasing function. Purchasing may also receive information on quality methods and standards, which must be understood and communicated to current and potential suppliers.

Purchasing should communicate regularly with the finance and accounting areas regarding forecasting of future needs, making major purchases, financing capital expenditures, tax issues, and long-range planning. As discussed in chapters one and two, purchasing must be actively involved in the organization's strategic and operating plans. Such involvement is important because purchasing may have specialized knowledge to contribute, and because purchasing will be held accountable for the achievement of these plans. It is important that purchasing's goals and objectives are integrated with those of the organization as a whole, and with other functional areas and divisions.

Because purchasing communicates with a wide variety of functions and levels within the organization, members of the purchasing function must be sensitive to the needs of each party. Vertical interaction represents members of a function communicating with others in the same function, such as a buyer to a purchasing manager. In addition to this intrafunctional communication, there are two basic types of interfunctional interactions: horizontal and diagonal. In the example of horizontal communication illustrated in Figure 4-1, the purchasing function is interacting with a party from a different functional area that is at a similar job level within the organization. This could also be a buyer dealing with an engineer, or the vice president of purchasing interacting with the vice president of marketing. While each represents a different function, they have similar interests regarding the level of detail they want in order to do their job effectively.

However, for the purchasing function, most of the communication is diagonal communication which occurs between different functions and different levels. Thus, it is important for members of the purchasing function to be aware not only of the type of information another person wants, but also the likely level of detail. For example, in dealing with a receiving clerk, purchasing should be able to supply details regarding purchase order numbers, carrier, items ordered, time of scheduled arrival, and so on. But if the director of distribution contacted purchasing to request information, the director would probably be more interested in the bigger picture, such as how much material was arriving and when. It is the responsibility of the buyer to determine the information needs of internal customers, and to provide those parties with information that is useful to them. In this way, purchasing can add value and gain the respect of other functional areas. If the purchasing function is insensitive to the type of information and the level of detail needed by internal customers, they will be more likely to bypass the purchasing function and delay its involvement in key issues.

FIGURE 4-1
Types of Organizational Interaction

Horizontal Communication

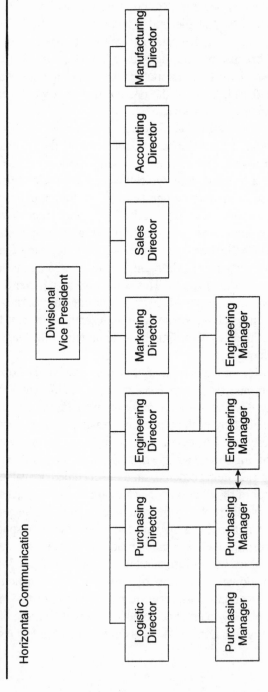

Shown: horizontal communication between purchasing manager and engineering manager

FIGURE 4-1 Continued

Diagonal Communication

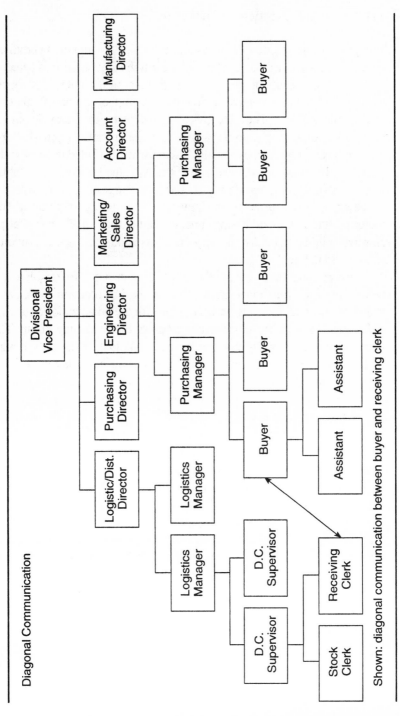

Shown: diagonal communication between buyer and receiving clerk

The Role and Perception of Purchasing

The purchasing function's actions and its ability to support internal customers largely determine the way that the purchasing function is perceived within the organization. This perception governs the interaction of purchasing with others, thereby having a direct impact on the role that purchasing plays. Some people feel that "anyone can buy," so they do not view the role of purchasing as important. In fact, some people regularly bypass the purchasing function, essentially buying what is needed to support their activities. Purchasing must show that it adds value by locating, establishing, and maintaining a good supply base; by suggesting alternatives; and by being proactive in problem situations. The more proactive and "customer-oriented" the purchasing function is, the more likely it is that purchasing will be included in important decisions and looked upon as strategic in nature.

At the same time, it is not the role of purchasing to follow blindly the wishes of other functional areas. The purchasing function must have a broad perspective, focusing on what is best for the organization as a whole, rather than pleasing the engineering group or the marketing group. Thus, purchasing must be well-informed and able to hold its own under pressure. In conflicts among internal functions, purchasing may play the role of a neutral third party, highlighting the strengths and weaknesses of each position in order to establish the consensus or compromise needed to move ahead. Playing such a role is only possible if purchasing has credibility with the functional areas involved.

Aetna Life and Casualty recently implemented a streamlined purchasing and payment process, aimed at focusing purchasing efforts on value-added activities rather than "order placing." As part of that process, authorized users can order goods and services directly from their computer terminals, drawing on national contracts prenegotiated by the purchasing department. Aetna's purchasing department focuses strongly improving internal relationships. The department has added an "800" number, which can be used by anyone who would like quick access to purchasing. The goal is providing outstanding customer service.[1]

On the other hand, purchasing may also be called upon to resolve conflicts and issues among internal customers and suppliers. Again, the

---

[1] Evans-Correia, Kate. "Aetna puts CAPP on Uncontrolled Buying." *Purchasing,* May 20, 1993, pp. 69-70.

role of purchasing is to be a neutral party—to look realistically at both sides of the issue and work toward arriving at a fair resolution that represents the interests of both the supplier and the internal customer. Fairness to supplier interests is critical, because purchasing is generally the key interface with the supplier base. Purchasing may have worked long and hard in selecting and maintaining a good working relationship with the supplier. Taking a position in support of the supplier may create conflict between purchasing and internal customers, who probably believe that purchasing should be representing their interests. Yet if purchasing is to continue working with that supplier, it must maintain a good working relationship and a certain level of trust.

In general, the amount of authority and responsibility that the purchasing function has is largely determined by how successfully purchasing performs its duties, and how proactive the purchasing function is in pursuing new opportunities to support internal customers, develop suppliers, and save the organization money. The purchasing function is responsible for educating its internal customers regarding its role, and how purchasing can support each internal customer group while supporting the objectives of the organization as a whole.

Establishing Credibility and Trust

As mentioned previously, it is up to the purchasing function to establish credibility and earn the trust and respect of internal customers. Members of the purchasing function can establish or reinforce such credibility by performing their duties in a professional manner, on a timely basis, and in a way that is responsive to the user's needs while keeping overall company goals in mind. After internal customers are convinced that the purchasing function's involvement in activities adds value, they will begin to include the purchasing function in pertinent activities on a regular basis. Purchasing must establish itself as a team player. This will reinforce its involvement and contribution, so the purchasing function can continue to build momentum toward being regarded as a key part of achieving the organization's goals.

Relationships With Other Functional Areas

Which functions purchasing interacts with depends on the nature of the organization and the activities the organization is involved in. This section discusses some of the key functions and the nature of the information that is, or should be, exchanged. The discussion first centers on functions common to all types of organizations before moving to functions that are

FIGURE 4-2
Overview on Internal Information Flows from Purchasing

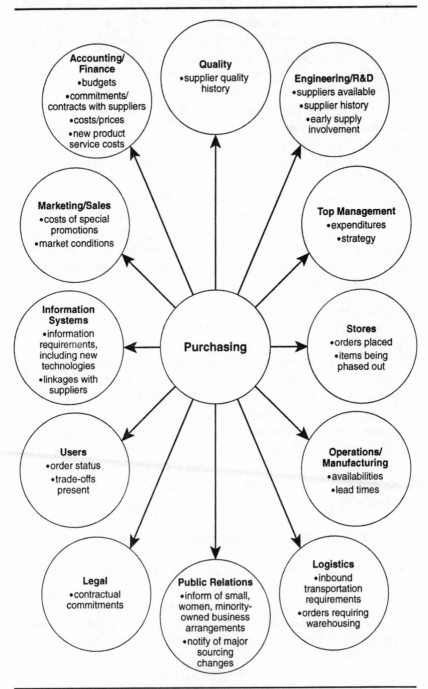

specific to manufacturing, non-manufacturing, or public organizations. A diagram depicting typical internal information flows from purchasing is shown in Figure 4-2.

Top Management

Purchasing dollars represent a significant amount of the total expenditures of an organization. Purchases of goods and services account for over 65% of expenditures in manufacturing organizations, nearly 86% of expenditures for wholesalers, 25% for service, and over 30% for the government.[2] Thus, top management has a strong interest in purchasing activity as a major use of funds and a major cost savings opportunity.

The top purchasing person reports to top management, either directly or through another function such as manufacturing or finance. A strong, progressive person is needed in the top level purchasing position, in order to enhance the function's status and visibility in the organization. The top level person should emphasize purchasing's contribution to the organization. The attitude of top management toward the purchasing function is the major determinant of how it will be viewed by the rest of the organization. If the purchasing function is viewed as a low-level, administrative function rather than a strategic contributor to the organization's success, purchasing faces the danger of being "squeezed out," as other functions take on the more strategic purchasing activities. Thus, the understanding and support of top management is critical to the purchasing function.

Engineering/Design/R&D

Many manufacturing and non-manufacturing organizations contain one or more functions that are heavily involved in research and design of new materials, products, capital equipment, and/or services. Decisions that affect the design of new or existing products/services can have a major impact upon the purchasing function's ability to perform its job. It has often been stated that 80% of product cost and quality is determined during the product design phase. Yet too often purchasing is not brought into the decision until after a supplier has been identified—after it is too late for purchasing to add much value.

In order to get around the problem of being involved "after the fact," a few things must be done. First, the company should establish a policy that all supplier contacts within the organization are to be initiated by the

---

[2] Herberling, Michael E. "The Rediscovery of Modern Purchasing." *International Journal of Purchasing and Materials Management,* Volume 29, No. 4, Fall 1993, pp. 48-53.

purchasing function. This should eliminate or greatly reduce "back door" selling. This is not to say that purchasing needs to be present at all supplier meetings. However, in order for purchasing to perform its job effectively, the purchasing function must be aware of new projects, and of the suppliers that are candidates for those projects. Many organizations have formal "supplier policy" statements to this effect, which are distributed to all current and potential suppliers. An example, taken from Bristol-Myers Squibb's "Suppliers Guide to Purchasing," is shown in Figure 4-3. It indicates that purchasing is the primary contact with suppliers, and that purchasing should arrange or be informed of all supplier contact with other functions. Such a policy helps prevent back door selling.

Purchasing may also be able to contribute by identifying other potential suppliers, and by pointing out the past performance of potential suppliers. To encourage full participation by the purchasing function in new product/service offerings or modification of existing product/service offerings, many companies have formally instituted early purchasing involvement. Under such an arrangement, purchasing is an active member of a development team, playing a proactive role in identifying and qualifying suppliers. Early involvement can reduce new product/service time to market by creating a capable supply base early in the project.

As an early participant, purchasing will become aware of unique design or performance requirements that are sensitive and need greater attention. Thus quality and delivery may be improved. Purchasing may also be in a position to suggest standard or lower-cost materials, which can reduce costs and time-to-market. Such new product/service knowledge and involvement allows the purchasing function to better serve internal users, adding value. As previously discussed, when the purchasing function demonstrates that it adds value, purchasing is more likely to be asked to participate in other important decisions, making the jobs of all members of the development team easier.

Whether the organization is in a manufacturing, non-manufacturing, or public sector environment, purchasing needs to be involved very early in any projects that change or develop new product or service configurations. Such changes may have a major effect on what is being purchased, and may represent excellent opportunities for the purchasing function to contribute via standardization, cost savings, cycle time reduction, and supplier identification and qualification.

Quality Assurance
Many organizations have eliminated separate quality assurance departments, making quality the responsibility of all functions within the organization,

FIGURE 4-3
Example of Supplier Policy Statement

General Information for Our Suppliers

- If a sales visit is planned, we highly recommend scheduling an appointment to make your meeting with us more productive. Our normal office hours are 7:30 a.m.-4:00 p.m., Monday through Friday.

- Upon entering any of our facilities, all visitors are required to register with a receptionist and obtain a Visitor's Badge. The receptionist will contact the party you are visiting.

- Beyond the reception areas all visitors must have an escort and must observe security and safety regulations at all times.

- Supplier contacts with other departments are generally arranged through the Purchasing office; however, direct contacts are permissible and appropriate, subject to Purchasing Department approval.

- It should be clearly understood that all purchase commitments for goods and services are the sole responsibility of authorized buyers within Bristol-Myers Squibb. Accordingly, we ask that all discussions dealing with the business aspects of our relationship either include a purchasing representative or be communicated to your purchasing contact in a timely manner.

- It is the policy of Bristol-Myers Squibb to treat all suppliers fairly. We expect the same professional conduct from our suppliers.

- When possible, purchases are made on a competitive-bid basis; however, your commitment to quality and service is of utmost importance.

- We appreciate your kindness; however, we are not allowed to accept business gifts from our suppliers. Your cooperation with this policy will be appreciated.

Minority Suppliers

- Bristol-Myers Squibb is a member of and subscribes to the goals and objectives of the National Minority Supplier Development Council. In addition, the Bristol-Myers Squibb, Mead Johnson Nutritional Group is a member of and supports the Tri-State Minority Supplier Development Council, a local organization providing regional support to area minority businesses. Within our organization we have established an active and aggressive small and small-disadvantaged business utilization program to identify, support, and increase our purchases from minority businesses.

Environmental Stewardship

- Without compromising the quality and competitiveness of our products, Purchasing, in partnership with our suppliers, is committed to minimizing any adverse environmental impact directly associated with materials and services purchased by our Company.

Source: Bristol-Myers Squibb, Mead Johnson Nutritional Group, used with permission.

such as purchasing, users, and engineering. In organizations where a quality assurance/quality control department still exists, it is very important for the purchasing function to establish a good interface. Purchasing works closely with quality in establishing the specifications and passing them along to suppliers. Quality must communicate well with the purchasing function regarding incoming quality from suppliers, so that purchasing is aware of supplier performance and can provide suppliers with feedback.

A very important interface between purchasing and quality exists during the supplier quality assessment, supplier selection, and certification processes. Quality personnel should work closely with purchasing in evaluating the supplier's capability to meet specifications. This includes active involvement in the on-site supplier assessment. Quality has the expertise to identify potential problems and suggest solutions, thus helping suppliers improve their quality. In addition, if the organization has a supplier certification process, the quality group will be a key player.

Accounting/Finance

Purchasing interfaces with the accounting/finance function in several key areas. Accounts payable pays the suppliers. Late payments can create a major strain on buyer-seller relationships. Thus purchasing must be aware of company policies and practices regarding supplier payment. If the organization pays in 40 days, purchasing should have that information incorporated into the terms of sale, so the supplier knows what to expect. Perhaps there are some cases, with small or highly leveraged suppliers, in which arrangements can be made to pay the supplier earlier than normal policy would dictate.

If purchasing is required to approve invoices for payment, it must get the invoices to accounts payable on a timely basis. This allows accounts payable the opportunity to take advantage of applicable discounts, and to better plan for upcoming cash outflows.

When purchasing negotiates major contracts, it must inform the finance area, so that finance can plan for the funds needed and make sure that the terms are acceptable to the organization. This is especially important if the contracts are long-term, if they require some type of regular or progress payments, or if they involve capital equipment or real estate purchases, which tend to be very large.

The accounting/finance function in many organizations administers the planning process. Purchasing may provide accounting/finance with a capital spending budget as well as a departmental budget to cover administrative expenses. In addition, particularly if a company uses standard costs, purchasing must provide accounting with price estimates for purchased goods and services for planning purposes. Such estimates are often updated quarterly for major expenditure categories, as the organization compares actual performance to plan and attempts to estimate earnings for the remainder of its fiscal year. When a new product or service is introduced, purchasing should provide accounting/finance with estimated costs of inputs, so that estimates of the product/service cost and associated profitability can be developed. Thus purchasing may work side-by-side with accounting/finance on a new product/service development team.

Marketing and Sales

Excellent two-way communication with marketing is critical to support effective buying. Marketing/sales often conducts special sales that can increase product/service demand dramatically. To support increased demand, increased purchases are required. If purchasing is aware of demand peaks and valleys in advance, it can shop for the best prices and keep suppliers informed of the organization's needs, so that expediting and shortages are prevented. Without such communications, the organization's total costs can rise dramatically. If special needs exist, such as a unique promotional package, sales promotion materials, a different size package and so on, purchasing should be informed as early as possible to prevent supply problems. Purchasing should also inform marketing of the additional costs associated with special packaging requirements, so that marketing can assess the true profitability of special promotions. Marketing/sales should also keep purchasing informed when product and service offerings are being phased out, in order to prevent excessive inventory build-up.

Information Systems

Many large organizations have an internal department called management information systems (MIS) or something similar. This department creates, updates and coordinates the organization's computer resources. With the increased availability, speed, and features of computer systems today, much of what purchasing does is becoming automated. Purchase orders are created and placed electronically, via electronic data interchange (EDI). Receiving is occurring via a computer terminal at Ford Motor Company, eliminating much of the manual document matching and checking of extended prices upon receiving incoming orders. In addition, no special invoice is needed, because Ford pays automatically based on purchase order quantity and prices. Orders that do not exactly match an open purchase order are refused by Ford's receiving personnel. This is an example of simultaneous internal interfaces between a variety of functions, including MIS, purchasing, accounting, and receiving.[3]

The Army and Air Force Exchange Service, a command of the Department of Defense, uses a sophisticated management information system in its post and base exchanges (stores for members of the military and their families). This retail point of sale (POS) inventory management system falls under the sales directorate, which combines the marketing and purchasing areas. The POS system tracks by item sales, prices, and markdowns, and is used for managing inventory and placing orders—some of which are resupplied directly by the supplier. Within the sales directorate, there is a sales information area, which provides the support for POS. This integration of purchasing, sales, and information systems helps keep the POS system operating smoothly.

It is imperative that the purchasing function have an excellent relationship with the MIS group. Purchasing should have an ongoing dialogue with MIS regarding new information technologies available, the information needs of purchasing, potential applications and problem areas, and related issues. This is important because without open communication, the needs of purchasing may not be fully examined and considered when system upgrades occur. As a result, purchasing could lose opportunities to upgrade and improve its systems. Today's improved computer systems reduce redundancy and manual, clerical type of work. Better systems can allow purchasing to focus on more strategic issues.

The User Community

Every buyer needs to have empathy and good communications with the user of the commodity or service that the buyer purchases. The user community has already been discussed as it relates to particular functional areas. But users vary widely, depending on the organization and the type of purchase. There are many types of "end users" that have not been mentioned. In a university, users may include professors and students. For the buyer of office supplies, everyone in the company is a customer. It would probably not be beneficial for the buyer to speak personally with each end user in such situations. However, it is important to be responsive to inputs from users, and to consider their perspectives in making the buying decision. Perhaps the secretaries or administrators who coordinate the office

---

[3] Hammer, Michael, and James Champy. *Reengineering the Corporation.* New York: Harper Collins Publishers, 1993.

supply needs of a particular department could be called upon for input. Similarly, the buyer of corporate travel services or fleet services needs to understand the critical needs of the users. Is price a key factor, or should convenience and ease of service take precedence?

In determining such trade-offs, the buyer needs to understand the goals of the organization as well as the issues facing the users. In the public sector, the "public" is the ultimate user or customer. The public purchaser may need to address a very broad user community. Receiving input from a group of users during the supplier selection process and receiving subsequent feedback regarding actual performance is very important. One must remember that part of what the purchasing function does is provide a service to the organization and to other functions within the organization. If purchasing does not add value and is not sensitive to user needs as well as the organization's goals, it will be bypassed in both major and minor decisions.

Legal

In acting on behalf of the organization as a purchasing "agent," purchasing has a great number of legal obligations and duties. In addition, purchase orders and other contracts are binding legal documents, created by the purchasing function on a daily basis. While most purchasers are aware of basic legal issues, purchasing should not hesitate to call upon the legal department when reviewing and establishing contracts. Some organizations even require that a member of the legal staff be present in major negotiations. As more organizations "outsource" functions, legal staff may be hired externally rather than internally. Further, many organizations do not need a full-time legal staff.

Public Relations/Public Affairs

Many organizations have a public relations/public affairs function to help establish and maintain a good corporate image in the community. Purchasing should keep public affairs informed concerning how the organization's money is spent in the community, particularly when small, women-owned, and minority-owned businesses are involved. If purchasing decides to move away from a local source with which the organization is currently doing a great deal of business, public affairs needs to be informed in order to anticipate any community backlash.

Stores/Facilities

The stores or facilities group is concerned with storing purchased items. Good two-way communication is required between stores and purchasing, in order to assure that space is available for incoming purchases. These groups may work together to establish stocking levels and order sizes, and to identify obsolete and excess material. Stores often requests goods, which creates purchase orders, so purchasing and stores must communicate regarding any items that are being phased out or experiencing lower or higher volume. Order sizes can then be adjusted accordingly. Non-manufacturing and public organizations may have a stores function for office and computer supplies, spares, or to manage other inventory needs.

Operations/Manufacturing

Operations/manufacturing functions exist in some form in virtually every manufacturing organization. These functions may exist in non-manufacturing and public sector operations as well, in order to manage the creation of the organization's services. Purchasing must have an excellent working relationship with the operations/manufacturing area of the organization. It is critical that the purchasing function receive realistic due dates and quantity requirements from manufacturing. If manufacturing does not trust purchasing, it will order excessive quantities and create artificial "rush" situations by requesting earlier due dates than what would otherwise be required. Such action costs the organization money in terms of excess inventory and higher obsolescence risk, as well as higher storage and handling costs. Artificially early due dates can create the need for expediting and premium transportation. They can result in higher storage charges, because the inventory is held longer. In addition, excess and early orders reduce the organization's available working capital, which is tied up in inventory earlier and for longer periods of time.

In order to effectively perform its job, the purchasing function must educate manufacturing/operations on the evils of excess inventory and the costs of early due dates. Purchasing should provide manufacturing with reasonable lead times, so that manufacturing knows how far in advance orders must be placed. Purchasing must establish credibility with manufacturing by selecting suppliers that deliver goods on time, as ordered, so manufacturing does not need to give purchasing false information in order to ensure that its production or service needs will be met. Purchasing must also be educated regarding the seriousness of late or inaccurate shipments, and buyers must understand which items are so critical that their absence could create a shutdown or a major bottleneck. Purchasing must develop a reliable source of supply for those critical items.

Logistics

Logistics is a broad function that encompasses inbound and outbound transportation, warehousing, and distribution. Such a function may not

FIGURE 4-4
Linkage of Purchasing and Logistics at Intel Corporation

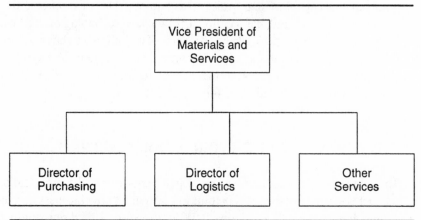

exist in some service organizations. In many organizations, logistics and purchasing are seen as so dependent that they are in the same reporting organization, which is often termed materials management. Such an organization, called corporate materials, exists at Intel, a major manufacturer of computer chips and personal computers (see Figure 4-4).

At Intel, both the director of purchasing and the director of logistics report to the vice president of materials. Both logistics and purchasing are working from a set of common goals and objectives, and each area understands how its activities affect the other function. The purchasing of outside transportation and warehousing services is really just purchasing another commodity, so all the purchasing requirements apply. Regardless of the organization, purchasing and logistics must coordinate their efforts. If purchasing is not directly responsible for inbound transportation, it must still coordinate with logistics in order to get transportation that meets the needs of the purchase. Such issues as the weight and density of the items, distance, lead times, and others should be considered. This allows logistics to match the transportation requirements with the organization's needs. Logistics and purchasing must also coordinate any storage requirements for the purchased items.

At a more strategic level, purchasing and logistics can reduce lead times and storage requirements by working closely together on understanding the organization's needs and supplier capabilities. They can use information technology with suppliers and carriers to speed transactions, to improve coordination and communication, and to drastically reduce paperwork and errors. In order for such an information system to be successful, purchasing and logistics must coordinate their information needs with supplier capability and available technology. Use of electronic data interchange (EDI) and bar coding are two key technologies that can improve purchasing and logistics efficiency. They allow the buying organization to communicate rapidly with carriers and suppliers, to coordinate shipments and receiving, and to manage internal inventories. These technologies will be discussed more fully in Chapter six.

INTERDEPARTMENTAL INTERACTION STYLES

To understand the way that various departments within the organization interact with one another, it is helpful to understand how business organizations have evolved over the past 100 years or so. This discussion will provide a background for understanding three major types of interaction patterns: functional silos, interactive, and teams or committees.

Evolution of Business Structure

Around the time of the founding of the United States, companies tended to be "one-person" operations. The companies were generally small and specialized, serving a localized region. One or few people controlled the entire operation.

Toward the middle of the nineteenth century, as companies began to grow, a few people could no longer manage all the organization's operations. Companies began hiring people to specialize in working with or managing various functions, such as manufacturing, sales, accounting, and so on. It was believed that this created efficiency and expertise. In many cases, the role of purchasing was retained by the owner of the organization.

By the turn of the century, as companies continued to grow and diversify their product/service offerings, functional specialization by itself was no longer enough. Large organizations began to divisionalize, organizing vertically around similar product/service offerings. Employees became specialized both in terms of function and product. Governmental organizations, including the military infrastructure, also expanded tremendously at that time. In some organizations, functions that did not directly affect the organization's product or service offering and that cut across divisional boundaries were left at a "corporate" level, supporting various divisions. This was common for functions such as human resources, accounts payable, purchasing, logistics, and treasury. There was no reporting relationship between "line" divisional employees, and "staff hr." or corporate employees.

By the 1950s, some large organizations realized that the divisionalized structure was not working well. It did not provide linkages between line people in various divisional and corporate positions, so the synergies of being part of a large corporation were lost. To combat this problem, many organizations began to implement a matrix. A matrix structure overlays the divisional structure, rather than replacing it. In addition to divisional reporting relationships, managers in a matrix organization have reporting responsibility to another person in their function outside of the division, often at a corporate level. This structure may also be used to create reporting relationships for special projects that straddle two or more divisions.

As we approach the end of the twentieth century, there is much speculation about the next prevailing form of organizations. Organizations have increased their outsourcing, contracting for many activities that were once done internally, so some observers speculate that a "hollow corporation" will develop. This hollow corporation, also called a network, will exist as a small organization of managers and "idea people," who hire external companies to perform all types of activities, including manufacturing, distribution, billing, and even sales and marketing.

The rationale for this type of organization, which represents a move away from diversified conglomerates, is that organizations should specialize and focus their efforts on what they do best, and should hire specialists to perform other activities. A variation on this is the concept of the "virtual corporation," in which a number of companies come together to develop, produce, and distribute/sell a product or a service of limited scope. These organizations establish a very close working relationship, which exists only while the product or service is viable. An organization may be simultaneously engaged in a number of such relationships involving a variety of products and services. These organizations focus on the product or service to be delivered to the customer, relying heavily on interorganizational and interfunctional teams. This type of organization is apparent in strategic alliances such as the relationship between Apple Computer, IBM, and Motorola, which was created in order to develop a comprehensive microprocessor and operating system for future generations of computers.[4]

---

[4] Davidow, William H. and Michael S. Malone. *The Virtual Corporation.* New York: Harper Collins Publishers, 1992.

Such relationships will be discussed in greater depth in Chapter 5. With the evolution of various forms in mind, the discussion now turns to how the various forms affect internal relationships.

Functional Silos

The phrase "functional silos" signifies the type of organization in which each functional area, such as purchasing, finance, marketing, and accounting, focuses primarily on the duties of its function, rather than the success of the corporation as a whole. The function may have defined its duty or mission with the overall corporate perspective in mind. However, it is not sensitive to the manner in which its activities affect and are affected by the efforts of other functions in supporting overall corporate objectives.

To use a simple illustration, the purchasing function may have the goal of providing the organization with the lowest-priced inputs that meet specifications. The manufacturing group may have the goal of providing timely, high-quality products. The distribution/logistics group may have the goal of getting the product to the customer on time, and in the most cost-effective manner.

In this example, by focusing primarily on the goal of low price, purchasing may choose a supplier with a delivery lead time that varies. The supplier's delivery may frequently arrive late or early. Thus, while purchasing is meeting its goal of low price, it may inadvertently sabotage the goals of manufacturing and logistics by creating "late" product. This in turn may create significant costs for the company as line stoppages occur, orders are expedited, and orders must be shipped to customers using expensive "overnight" methods. The savings accrued from using the low-price supplier may actually increase the company's total costs.

Therefore, while purchasing may believe that it is meeting its goals, the function is not really supporting the organization as a whole. By failing to examine the broader picture to understand how purchasing decisions affect other functional areas and the organization's overall efficiency, purchasing may actually be incurring higher costs.

A functional silo mentality or culture is very difficult to change. Each member of a functional area tends to develop loyalty and commitment to his or her function, with the needs of the total organization coming second. Interactions with other functions become a zero-sum proposition. Adversarial relationships may develop among functions within the organization, as they vie for scarce resources and strive to achieve goals that may be in conflict. The advantages of specialized functions, such as focus, expertise, and scale may be defeated by myopia and poor communication.

Interactive Relationships

To overcome some of the problems associated with compartmentalized functions, many companies have divisionalized. Each division may have a complete line and staff organization that works toward the goals of the division, with one or more specialists taking care of each of the functions. However, when divisions become large and there are large groups of specialists within the same functional area, functional silos may develop at the divisional level. Employees may spend more time interacting with others from their own function than with other functions. One way to get around this is to intersperse the offices of different functions, rather than having each functional group sit together. This is often referred to as "functional co-location."

In addition, some companies have used the previously discussed matrix organization, in which employees have dual reporting responsibility. For example, an employee might be a buyer who reports to a purchasing manager. However, that employee may also be assigned to a specific division or project. In this capacity, the employee may report to a manufacturing manager on the divisional level. This dual reporting responsibility is designed to help employees understand and become loyal to the entire business, as well as their functional area. The goal is to utilize employees' expertise for the good of the division, rather than using it only to preserve the function that the employee represents. The employees are closer to day-to-day operations, thus gaining a better understanding of how their decisions affect the organization as a whole. It may, however, be difficult for an employee to have two managers. This is particularly true if functional and divisional goals conflict.

The Team Movement

To counteract some of the problems of poor interfunctional communications, organizations are increasingly using work groups. The goal of a work group is to combine the skills and expertise of a number of people, generally from different functional areas, in order to develop a better plan, decision, or execution of some action than would develop if each person worked on the problem individually. These work groups can take on a number of forms, including ad hoc committees, standing committees, task teams, and work teams. The level of time commitment and ongoing responsibility varies, but ad hoc committees generally represent the lowest level, and work teams often represent the highest level. The differences among these various types of work groups is shown in Figure 4-5.

FIGURE 4-5
Types of Work Groups

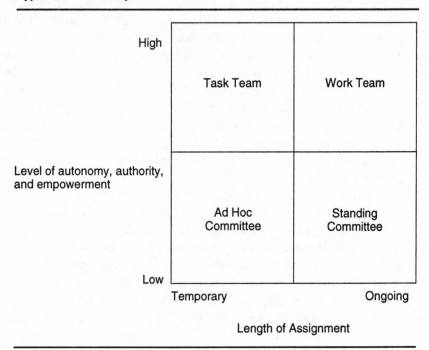

Ad Hoc Committees. Ad hoc committees are formed to resolve a certain, focused issue, generally within a time frame. For example, a committee may be formed to recommend improvements in how office supplies are ordered and managed. Once the committee has made its analysis and recommendation, it will probably disband, never to exist in that form again. Ad hoc committees vary in intensity and duration, depending on the type of decision to be made and the urgency or criticality of the decision. Ad hoc committee members may be very dedicated to the goal they are trying to achieve or the decision they are trying to make. They may not need to be dedicated to the other team members or the functioning of the team, because they know it is a temporary assignment. Thus they may not want to expend much effort on group dynamics issues. By definition, committees are generally not all encompassing; the employee has other duties to maintain while serving as a member of one or more committees.

Standing Committees. Standing committees are distinguished from ad hoc committees in that they are of indefinite or permanent duration. For example, there may be a committee in charge of supplier certification. As long as the organization continues to certify suppliers, that committee will exist in some form. A distinguishing feature of standing committees is that they generally need approval from top management or parties outside of the team to proceed with their activities.

Because standing committees are ongoing, their structure and function tends to take on more importance than is true with ad hoc committees. Southern Pacific Lines uses a standing committee approach to commodity management. A committee is established for commodities that require a high attention level, and this group includes representatives from all departments that have a vested interest in the commodity. The committee establishes and executes the commodity management plan, which is reviewed, approved, and monitored by a steering committee of top management.[5]

Teams

A team is defined as "a small number of people with complementary skills who are committed to a common purpose, set of performance goals, and approach for which they hold themselves mutually accountable."[6] A key difference between teams and committees relates to their accountability and the manner in which they actually perform work. The distinction is made here between "task teams" and "work teams." Task teams exist for a specific, identifiable purpose with a clear end. Work teams are ongoing in nature, much like a divisional structure, with specific, continuing goals.

Task Teams. While both task teams and ad hoc committees focus on a discrete task, they function differently. The key distinction is that the members of an ad hoc committee still function primarily as individuals, not as a team. Ad hoc committees may meet to decide what has to done, but then each person individually completes his or her "piece" and is individually accountable for the results. In a task team, there is mutual accountability, and much of the work may be performed with other team members.

---

[5] Berndt, Roger. "Commodity Team Management: A Cross-Functional Approach to Procurement." National Association of Purchasing Management Annual Conference Proceedings, 1993, Tempe, AZ: NAPM, pp. 45-49.

[6] Katzenbach, Jon R. and Douglas K. Smith. "The Discipline of Teams." *Harvard Business Review*, Volume 38, No. 2 (March/April 1993), pp. 111-20.

Another important distinction is that the task team "owns" the project it is working on. For example, a task team formed to implement supplier certification would be fully empowered to design and implement the program without further approval. However, an ad hoc committee would require approval from parties outside of the team, generally top management, before proceeding with each step. Thus a task team has a much higher level of task and group responsibility than an ad hoc committee.

Work Teams.   Returning to the definition of teams, work teams, also known as self-directed teams, are distinguished by their commitment to a common purpose and goals, but more importantly by their ongoing mutual accountability. These cross-functional groups are often organized around a product or a service offering, and may be responsible for all aspects of that product/service, from design and development to customer support.

The work team framework is unique, in that it is not a temporary structure that overlays another organizational form, as are other teams and committees. It should also be noted that the work team type of organization is relatively rare in practice. It represents a major change for most organizations, and it should only be pursued if benefits exceed costs. A key question to ask is whether joint performance and extensive cross-functional interaction will yield significantly better results than a more traditional approach. Only if the answer is yes should an organization implement work teams. Otherwise, committees may suffice.

One firm that has implemented a work team approach (empowered people on empowered teams) in several of its divisions is Motorola. In Motorola's government systems and technology group, four commodity teams with multiple technology sub-teams provide a work team structure. Each team is responsible for the engineering and purchasing aspects of a particular class of commodity, such as semiconductors and connectors. Teams are made up of materials quality engineers, one or more commodity engineering specialists, a price analyst, a planner, and a purchasing agent. Concurrent product development and cost analysis may also be included. These co-located teams are jointly responsible for the management of suppliers in terms of total cost, quality, and delivery of parts and services. The members are judged as a team, as well as on an individual basis.[7]

---

[7] Carter, Joseph R, *Purchasing: Continued Improvement Through Integration,* Homewood, IL: Business One Irwin, 1993.

Total empowered teams should not be confused with consensus management. Team members are empowered to make commitments on behalf of the team on many business issues. Consensus is used for systemic issues and problem resolution. The Japanese quality circles are based more on total consensus decision making.

Using Teams and Committees

Organizations are not limited to using one type of team and/or committee at a time. The structure used should fit both the activity to be performed and the culture of the organization. For example, Motorola may use the work team structure, while using committees and task teams that overlay the work team structure to perform activities that don't fit well with the work team concept. Because the work team structure is a new idea, organizations may evolve toward it slowly, using committees first, then task teams before evolving to work teams. Conversely, an organization may decide that work teams would not greatly benefit its performance.

Because most people have worked in environments where individual accountability was paramount, the use of teams and committees can be threatening and even dysfunctional. To combat potential problems, most organizations will train employees to function as a contributing group or team member. Training may include such topics as appreciating diversity/individual differences, team interaction, and team accountability, and may even involve team building exercises or activities outside of the workplace. Other issues that must be addressed in developing teams and committees are accountability, the degree of responsibility and decision making authority, and the impact of team/committee performance on individual performance appraisals.

No matter which formal structures are chosen by the organization to stimulate interaction, interfunctional relationships remain a key component of the purchasing job. Good working relationships with other functions are critical to purchasing's effectiveness.

KEY POINTS

1. Purchasing interacts with virtually every area within an organization, making communication a key issue.
2. The role of purchasing is to be proactive, and to make decisions that best support the overall goals of the corporation.

3. Establishing credibility and trust with other functional areas is key to the success of the purchasing function.

4. If purchasing is to be successful, it must understand the needs and goals of other functions.

5. Organizational structure is evolving from functional silos to more interactive relationships.

6. Team structures are growing in popularity, and purchasing often plays a key role in team leadership.

SUGGESTED READINGS

Burt, David N. *Proactive Procurement: The Key to Increased Profits, Productivity and Quality.* Englewood Cliffs: Prentice-Hall, 1984.

Burt, David N. and Michael Doyle. *The American Keiretsu.* Homewood: Business One Irwin, 1993.

Carter, Joseph R. *Purchasing: Continued Improvement Through Integration.* Homewood: Business One Irwin, 1993.

"Teaming: Bringing It All Together." *NAPM Insights,* March 1993, pp. 21-35.

CHAPTER 5

PURCHASING'S ROLE IN EXTERNAL RELATIONS

One of the key roles played by the purchasing function is representing the organization to outside organizations and individuals. This includes establishing the appropriate types of relationships with suppliers. The relationship of the purchasing function with suppliers is particularly critical, and is of growing importance to the success of the organization. Issues in supplier relationships include evaluation of how to establish good relationships; determining the organization's involvement in supplier education and training, supplier improvement, or supplier development efforts; and reciprocity, including supplier "partnership" or strategic alliances.

Purchasing also represents the organization to outside associations. These include trade associations as well as professional associations dedicated primarily to purchasing issues. The image that the purchasing function projects to outside associations affects not only the image of the firm, but also the image of the purchasing function itself.

Benefits of Good Supplier Relationships

The benefits of establishing and maintaining good supplier relationships are many. These benefits accrue directly to the purchasing function, and to the firm as a whole.

The performance of an organization's supplier base reflects directly on the purchasing function. Because purchasing is generally considered to have primary responsibility for the selection and ongoing management of the supply base, problems with suppliers must be managed by purchasing. The presence of ongoing supplier-related problems sends a message to internal customers that the purchasing function is not doing a good job.

If the problems are so severe that the end product or service visibly suffers, poor supplier relationships can make the whole organization look bad, and can hurt relationships with external customers.

Establishing good relationships with suppliers should mitigate problems for a number of reasons. First, when a firm has good relationships with suppliers, communication should improve. The supplier is more likely to communicate potential problems to the purchaser. The purchaser can then work with the supplier to solve or minimize these problems, or can find an alternate supply source if appropriate. Suppliers that are treated fairly are likely to be more responsive and cooperative in dealing with quality and service issues.

Regardless of the organization's operating environment, emergencies and special situations sometimes arise. If purchasing has established good working relationships, suppliers are much more likely to come to the organization's aid when requirements suddenly change. This makes the purchasing function look good, and it makes the job much easier from an administrative standpoint. In addition, as purchasing works toward reducing the organization's supply base, it has fewer suppliers to fall back on in case of problems. Thus it becomes critical to have good working relationships with the organization's remaining suppliers. This can be particularly important in times of shortage, temporary price fluctuations, or supplier problems.

Good supplier relationships are especially important in areas that are undergoing rapid technological change. Here, purchasing must rely upon the supply base to keep it abreast of technological developments, so the company's product or service offerings do not become obsolete.

A good supplier relationship, characterized by trust and recognition of mutual dependence, sets the stage for early supplier involvement (ESI) in new product/service design and development, or in modification of existing product/service offerings. Many organizations, including Pacific Bell, Chrysler Corporation, Motorola, the Arizona Department of Transportation, and MCI, involve their suppliers in the development of new products or services. In doing so, the buying organization utilizes the supplier's expertise and technical resources, and often develops better alternatives at a lower cost.

Good supplier relationships should also greatly improve the purchasing function's efficiency, for a number of reasons. First, by establishing a pool of good, reliable sources, the purchasing function should be able to reduce the amount of time and effort spent searching the supply market for potential sources. This should free-up the purchasing function to concentrate on more strategic issues. Second, purchasing will spend less time on unproductive issues such as expediting and identifying problems. The supplier, wishing to maintain good relations, will attempt to make the purchaser aware of potential problems in advance. Further, a stable supply base reduces uncertainty and allows for better planning of the firm's operations.

Good supplier relationships open the door for long-term agreements and the mutual understanding of each party's needs, which again increases the efficiency of purchasing. Good relationships are generally a prerequisite for supplier development and supplier partnering. Both of these approaches yield many benefits, which will be discussed later in this chapter.

Means of Promoting Good Supplier Relationships

Establishing and maintaining good supplier relationships is an active, ongoing process. It involves prompt payment, excellent communications, building trust, fair treatment, reducing the supplier base, being "easy" to do business with, and soliciting and responding to feedback from the supplier. For every obligation the purchaser has in creating good relationships, the supplier has a parallel duty.

Payment As Agreed
Ultimately, the essence of purchasing is the exchange of money for goods and services. As previously stated, mutual cooperation is very important in maintaining good supplier relationships. From the supplier's perspective, this means fulfilling contractual obligations as specified. From the purchaser's perspective, one of the key issues is prompt payment of supplier invoices.

It may seem obvious that paying suppliers on time is important. However, for control reasons, purchasing is rarely the function that actually pays the suppliers. Supplier payment is usually far removed from purchasing, occurring in the accounts payable department. Purchasing may have little say regarding payment terms, as the organization attempts to improve cash flow by delaying supplier payments. Nonetheless, it is up to the purchasing function to be aware of supplier payment policies, and to make suppliers fully aware of these policies so there are no "surprises." If the purchaser's organization, for whatever reason, continually pays suppliers late, the relationship between the organizations will likely deteriorate, and may come apart altogether. It may be impossible for purchasing to locate suppliers that provide high service levels if the purchaser's organization does not reciprocate.

Communications

Open, honest, two-way communication is the foundation of good supplier relationships. Much recent research indicates that poor communication is the key cause of failure in purchaser-seller relationships and strategic alliances, and is a contributing factor in quality problems. There are several elements of communication that are of concern to the purchaser. First, the purchaser needs to give the supplier an accurate assessment of the organization's needs and expectations over a reasonable time period. Just what constitutes a reasonable time period depends upon the supplier's lead times and the level of commitment required to meet the organization's demands. If problems occur in using the supplier's goods or services, or if the quantity, timing, or nature of demand changes, the buying firm should notify the supplier as soon as possible. The supplier can then make adjustments. This type of communication is critical. It opens the door for a dialogue that allows firms to plan around or mitigate problems.

Good communication from the buying organization encourages the supplier to keep the purchaser informed early of potential problems it may be experiencing, regarding production, its suppliers, quality, and so forth. Early advisement can in turn lead to supplier development or joint problem solving.

Trust

Trust is an important aspect of good supplier relationships. Trust requires good two-way communication, because each party needs to believe that it can rely on the statements and good intentions of the other. In addition, trust comes when both parties know they are being treated fairly and equitably. In supplier relationships, this is based on giving the supplier adequate time to respond to demands, fair consideration in competing for new business offered by the firm, and reasonable treatment and expectations. It relates to treating the supplier with respect, integrity, and consistency. If a supplier does not feel it has a fair chance of securing additional business because another supplier is favored or has already been chosen, or if it suspects that the purchaser is "going through the motions" of soliciting bids or proposals, then trust will be very difficult to establish.

Supply Base Reduction

Many organizations are reducing their supply bases. This makes it easier and more efficient for the purchasing function to select and manage supplier relationships. It also opens the door for improved relationships, because purchasing can work closely with the remaining suppliers. For organizations with very large supplier bases, the sheer number of suppliers makes it almost impossible to establish good relationships with any but the most critical suppliers.

Continuous Improvement

As discussed in more depth later in this chapter, both parties should work on continuous improvement of processes. This includes ordering methods, incorporating such technologies as EDI and barcoding, and service provided by the supplier to the purchaser.

Easy to Do Business With

No single characteristic creates the perception that an organization is "easy to do business with." Instead, the perception results from an overall attitude, or an approach to conducting business. An organization that is easy to do business with establishes good working relationships with its suppliers. It communicates well, treats suppliers fairly and equitably, does not ask for extra help, does not solicit bids unless they will be seriously considered, and pays suppliers promptly. It treats suppliers with respect and rewards them for "going the extra mile."

Solicit Feedback

How can a purchaser know whether his or her firm is "easy to do business with"? One way is by soliciting open, honest feedback from suppliers. Tennant Corporation has done this by sending its suppliers a "Supplier Feedback" form. This form, returned directly to Tennant, asks the supplier for feedback in a number of areas, as shown in Figure 5-1. Ford of Australia has used a similar format to solicit supplier feedback. However, the suppliers were asked to return the Ford forms to a third party. The goal was to assure anonymity, so the suppliers would feel free to be candid. Another approach, taken by Intel, is to have an open door policy. In Intel's supplier guide, the director of corporate purchasing states that, "Should we fail to live up to the standards described in this guide, I encourage you to contact me directly for a confidential review of the circumstances."[1]

In addition, an organization may hire third parties to conduct an anonymous survey measuring supplier attitudes toward the company and its competitors. This benchmark allows an organization to assess its relative strengths and weaknesses as compared with the industry. It can provide a more meaningful basis of comparison than a survey aimed only at the purchasing organization.

---

[1] *Supplier Guide.* Chandler, AZ: Intel Corporation.

FIGURE 5-1

Satisfaction Rating: Major Categories Surveyed by Tennant

    I.   Overall Business Relationship
   II.   Relationship with Procurement Specialists
  III.   Relationship with Engineering Personnel
  IV.   Relationship with Material Controllers
   V.   Supplier Quality Updates
  VI.   Relationship with Quality Assurance and Inspection
 VII.   Type of Training Support Tennant Should Provide
VIII.   Tennant's Greatest Strengths and Weaknesses in Doing Business
       with Your Firm

Source: Tennant Corp. Used with permission.

SUPPLIER EDUCATION, INVOLVEMENT, AND DEVELOPMENT

One potential goal of establishing good supplier relationships is to work toward improving the supplier's performance, or the supplier's involvement with the organization. This can be accomplished through supplier education; supplier involvement in design, development, and value-analysis; and supplier development, also known as reverse marketing. As organizations reduce their supplier base, they become increasingly reliant upon the remaining suppliers. Due to this greater dependence, there seems to be an increasing trend toward using all of these supplier improvement/involvement activities. Each technique is discussed in the following section.

Supplier Education and Training

Education and training of employees is a large expense for many firms. The concept of increasing the training program to include outside suppliers may seem cost-prohibitive. However, an increasing number of firms are doing this. These organizations view the expense as an investment that will more than pay for itself in terms of improved supplier performance.

One firm that has undergone a limited-scale but through joint purchaser-supplier training program is Pacific Bell in San Ramon, California. Pacific Bell worked with an outside group to develop an extensive quality training program, which was attended by its purchasers and a select group of suppliers. This 45-hour program was conducted during business hours, but it required a great deal of work outside of class. College credits were granted, and each successful participant received a certificate.

Other, less intensive training programs are also growing. Tennant Corporation has trained hundreds of its suppliers' employees in quality methods at Tennant's in-house seminars. Companies like Corning, Honda, and Motorola have invited select suppliers to attend their in-house training. The idea is that, if a supplier represents a long-term relationship, improving that supplier's skill level should provide long-term benefits for the buying firm.

Supplier Involvement

While early supplier involvement, or ESI, is a "buzz" term today, supplier involvement can exist on many levels. The supplier may be involved in the initial design, development, or redesign of a product, process, or service; or it may be involved in the value-analysis of an existing product for potential incremental change. Regardless of how the supplier is involved, the goal is the same: to utilize the supplier's expertise in improving the value and producibility of the item in question. There are many other benefits as well.

Early Supplier Involvement
Early supplier involvement is also sometimes referred to as concurrent engineering. This concept has been used successfully by many organizations, including Hewlett-Packard (printers), the Arizona Department of Transportation (information systems), and MCI Communications (data cards). ESI has helped these organizations and many others define better solutions for their problems, increase their competitiveness, reduce costs, and speed their cycle time to market.

ESI involves getting a potential supplier—or even two competing suppliers—involved very early in new product or service development; the earlier the better, in terms of the supplier's potential contribution. ESI usually involves a development team that includes the purchasing function, designers, and user group or engineering function, depending on the type of organization. Because of the sensitive nature of many new product offerings, the supplier is generally asked to sign a confidentiality agreement before the buying organization discloses any information.

If the supplier is involved at the conception stage, the buying organization informs the supplier of the type of service or product it plans to offer. The potential suppliers then work, either independently or with the buying organization's development team, to develop a cost-effective, functional method of meeting the organization's needs. This type of ESI parallels the concept of concurrent engineering. Concurrent engineering is one type of ESI.

Due to the novelty of the concept and the discomfort that many organizations feel about involving the suppliers in the conception of the product, more firms tend to involve the supplier at the design stage. Here, the firm has a much clearer idea of its potential product or service offering, which somewhat limits the supplier's opportunity to be creative. However, this approach yields significant benefits.

Involving suppliers in conception and design also helps develop supplier and purchaser processes. As the buying and supplying organizations develop an understanding of each other's needs and processes, they can work to ensure that the processes are compatible. The party who can most efficiently perform certain functions can take on those functions. ESI helps assure that the product or service developed will be producible and procurable, based on the constraints faced by both the buying and supplying organization, before the design is completely committed. These constraints include cost, technology, capacity, space availability, and related issues.

If the service or product will require a capital investment, early purchasing involvement (EPI), coupled with ESI can expedite the time-consuming capital appropriations and capital budget approval process. This in turn can speed up the acquisition and approval processes, reducing cycle time for purchase of the new asset. Earlier availability of the capital asset can provide the firm with the competitive advantage of earlier new product/service introduction.

ESI allows the organization to be more aware of the potential cost of purchased inputs. Early knowledge of potential cost overruns allows for change of design before the process is committed, in order to make the product or service cost-competitive. Thus ESI can prevent the development of products or services that are too costly to command a reasonable sales volume. Cost reduction may also be possible as the supplier works to meet the purchaser's needs and produce the required items efficiently.

Similarly, quality may be improved by incorporating what suppliers do best, and by considering interaction among various components early in the design process. Further, suppliers can institute the required quality methods or improvements in advance, to prepare for introduction of the new product or service.

ESI also improves product/service availability, because the supplier is involved in its specification. Such involvement should assure that the product or service is producible, and that the supplier has the necessary capacity.

Often, the supplier will utilize its expertise and knowledge of related applications to develop a completely new idea that the buying firm had never considered. This is yet another advantage of getting the potential suppliers involved as early as possible. In addition, by utilizing the supplier's technical resources, the organization may be able to reduce the number of technical people on its staff, reducing overall costs. ESI also assures that the product/component or service needed will be available at a reasonable cost, because the supplier is involved in creating the item, and is generally aware of cost constraints.

ESI should create a superior product or service by utilizing the synergies of the development staff at both the purchaser and supplier organizations. The supplier may be developing a new technology that the purchaser may be able to incorporate into the new product or service. Such a benefit would probably not be possible if the buying organization approached the supplier with a firm design in mind.

ESI should reduce the time needed to get the new product or service out to market, because a team has been working on producibility and availability issues all along. Reduced time to market is a key competitive issue for the 1990s, and is one of the key reasons for ESI. ESI should also help reduce the cycle time from order placement to receipt, because the organizations have worked together to streamline the purchasing process. Finally, ESI strengthens the organization's relationship with the suppliers involved, and may lead to a strategic alliance relationship.

Supplier Involvement in Redesign

An organization may choose to redesign a product or service in order to improve producibility, to update the product/service, or to add features. Generally, when enhancing an existing product or service, an organization will rely on existing suppliers. These suppliers are good sources because they are familiar with the product or service, and because they know the issues they have faced in trying to meet the buying firm's needs. Getting these suppliers involved early in changes can yield many of the same results as ESI, such as lower cost, improved producibility, and reduced cycle time needed to implement the change. Purchasing should be involved as a key liaison on the redesign team. Involving suppliers and purchasing early in the redesign process is a good first step for firms that would like to increase supplier involvement, but are not ready for ESI.

Supplier Involvement in Value Analysis

The development of the value analysis concept is credited to General Electric Company in 1948. Value analysis is commonly defined as a method for evaluating the functionality of a purchased good or service, and evaluating whether there is a more cost-effective way to achieve the same result. GE uses three steps: analytical, alternative generating, and implementation. There may be opportunities for supplier involvement in evaluating functionality, even beyond the items that it supplies directly. Such participation can yield synergies and create new ideas. One caution in eliciting supplier involvement is this: a supplier is not likely to suggest a way to design its product or service out, even if that is the best alternative. Value analysis is discussed in much greater depth in Volume Three of this series.

Supplier Development/Reverse Marketing

Like education and training, supplier development and reverse marketing reflect the organization's reliance on reduced supply base, and its willingness to invest in improving supplier performance. *The 1992 NAPM Dictionary* defines supplier development as:

> A systematic organizational effort to create and maintain a network of competent suppliers and to improve various supplier capabilities that are necessary for the buying organization to meet its increasing competitive challenges.

This broad definition clearly includes supplier education and training programs as discussed in previous sections. In addition, it encompasses the organization's direct efforts to improve supplier processes, quality, and other performance-related issues. The better the organization's relationship with the supplier, the more likely the supplier is to fully cooperate with those efforts. As the organization works to improve the supplier's performance for its own benefit, the supplier will also improve its performance to its other customers, making it a more competitive supplier overall. Thus the benefits are mutual. One firm that becomes heavily involved in supplier development activities is Honda of America. If one of its suppliers is experiencing a problem and is unable to solve the problem on its own, Honda sends in its own employees to help identify, isolate, and solve the problem, so the supplier can get back up to speed.

Closely related to supplier development is the concept of reverse marketing. Honda of America and NUMMI both apply this concept.[2]

---

[2] Newman, Richard G. and R. Anthony Rehee. "A Case Study of NUMMI and its Suppliers." *Journal of Purchasing and Materials Management,* Volume 26, No. 4, pp. 15- 20.

Supplier development, also know as reverse marketing, is appropriate for working with suppliers when sources are currently available, as well as for creating suppliers when no known source exists.

Continuous Improvement

At one time, an organization would tell its suppliers what was expected of them in terms of cost, quality, delivery and so on. Those expectations would remain static until there was a product/service or process change. That is no longer the case. Continuous improvement is another important concept that organizations are embracing in the 1990s, in order to survive and remain competitive. Just as an organization focuses on continually improving its own operations, it expects its suppliers to continuously improve. In order to fairly assess whether a supplier is improving, the buying firm must have a valid method for measuring the supplier's performance within various parameters.

The most common areas of focus for continuous improvement are cost, quality, and design. Other issues could include delivery, responsiveness, and innovation. Supplier performance measurement is beyond the scope of this book, but it is discussed in some detail in Volumes One and Two of this series.

Some organizations make continuous improvement expectations an inherent part of their supplier certification process. Certification involves an investigation of the supplier's facilities, processes, and output for quality levels. For example, Honeywell's Air Transport Systems Division issues its supplier certification guide with blanks for all of the performance expectation figures. Each supplier is then given a goal, including performance expectations for "Q1-Q3" and "D1-D3." These represent quality levels and delivery standards required to meet Honeywell's supplier performance classifications, as shown in Figure 5-2. Those figures are periodically updated to reflect Honeywell's continuous improvement goals. Likewise, Tennant Corporation, as part of its supplier certification, asks suppliers for new performance goals each quarter. These performance goals are to reflect the supplier's continuous improvement efforts.

Another form of continuous supplier improvement practiced by aggressive firms such as General Electric is supplier involvement in cost/design analysis. This can occur by inviting suppliers to become part of a value analysis team for existing products or services, or through ESI on new products and services.

FIGURE 5-2
Honeywell's Supplier Performance Expectations

Requirements for each level of classification are as follows:

Certified supplier

- QPR of Q1 ppm or below.
- DPR of D1 % or above.
- Quality Improvement Plan in place.
- Supplier Self Survey satisfactorily completed annually.
- Supplier Quality System Audit satisfactorily completed biannually.

Preferred supplier

- QPR of Q2 ppm or below.
- DPR of D2 % or above.
- Quality Improvement Plan in place.
- Supplier Self Survey satisfactorily completed annually.

Conditional supplier

- QPR of Q3 ppm or below.
- DPR of D3 % or above.
- Quality Improvement Plan in place.
- Supplier Self Survey satisfactorily completed annually.

Restricted supplier

- Supplier Self Survey satisfactorily completed annually.

QPR - Quality performance rating
DPR - Delivery performance rating

Source: Supplier Performance Classification System, Honeywell Air Transport Systems Division. Used with permission.

SUPPLIER ALLIANCES/PARTNERSHIPS/PREFERRED SUPPLIERS

Supplier alliances/partnerships have become one of the hottest topics in interfirm relationships. But despite all the interest in alliances, there is still a great deal of confusion about what constitutes an alliance, and when alliances make the most sense. This section will examine several perspectives on classifying strategic alliances/partnerships, then discuss how these relationships develop and evolve.

There are countless definitions for strategic alliances or supplier partnerships in use today. The *1992 NAPM Dictionary* defines a supplier partnership as follows:

A supplier partnership between a purchasing and supplying firm involves a mutual commitment over an extended to indefinite time horizon to work together to the mutual benefit of both parties, sharing relevant information and the risks and rewards of the relationship. These relationships require a clear understanding of expectations, open communication and information exchange, mutual trust, and a common direction for the future.

While this definition provides a good starting point for understanding the partnering concept, it is important to note that this is a very broad definition. In reality, there are many different types of partnering relationships, and many terms are used interchangeably to describe such relationships. Some of the terms in use today are: preferred supplier, alliance, strategic alliance, partnership, collaboration, cooperative agreement, American Keiretsu, and reverse marketing. These all represent some slant on the partnering theme. Often, these terms are used inconsistently. Thus the best approach is to clearly define the parameters of partnering, or whatever term is used, upfront. It is also important to note here that many organizations are moving away from the use of the term "partner," particularly regulated organizations such as utilities, because it implies a legal business entity.

Types of Partnerships

The continuum of partnering relationships is shown in Figure 5-3. Several demarcations are made in the middle of the continuum. These simply illustrate various ways of classifying partnering relationships. In reality, there is no clear break between one form of partnering relationship and another. This section will first discuss an approach to partnering suggested by SEMATECH, then explore another simple model.

SEMATECH Approach
The SEMATECH approach to partnering discusses the two extreme ends of the continuum: the basic partnering arrangement and the expanded partnering arrangement. Basic partnering is congruent with the idea of "good supplier relationships," in that an organization can have this type of relationship with all of its suppliers. However, SEMATECH goes deeper by stating that partnering per se is really a philosophy. As such, the basic partnering concepts apply to customers and employees of the organization, as well as to suppliers. The key characteristics of partnering are shown in Figure 5-4. Basic partnering embodies the idea of treating everyone with respect, honesty, trust, and integrity, and keeping the lines of communication open.

FIGURE 5-3
Types of Purchaser/Supplier Relationships

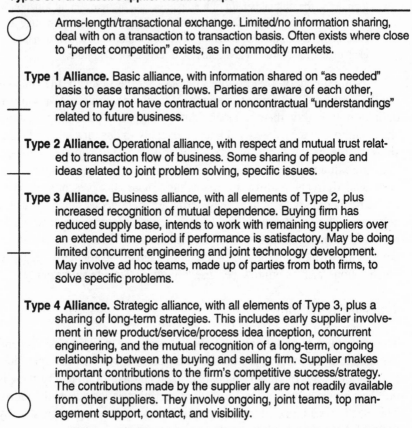

Arms-length/transactional exchange. Limited/no information sharing, deal with on a transaction to transaction basis. Often exists where close to "perfect competition" exists, as in commodity markets.

Type 1 Alliance. Basic alliance, with information shared on "as needed" basis to ease transaction flows. Parties are aware of each other, may or may not have contractual or noncontractual "understandings" related to future business.

Type 2 Alliance. Operational alliance, with respect and mutual trust related to transaction flow of business. Some sharing of people and ideas related to joint problem solving, specific issues.

Type 3 Alliance. Business alliance, with all elements of Type 2, plus increased recognition of mutual dependence. Buying firm has reduced supply base, intends to work with remaining suppliers over an extended time period if performance is satisfactory. May be doing limited concurrent engineering and joint technology development. May involve ad hoc teams, made up of parties from both firms, to solve specific problems.

Type 4 Alliance. Strategic alliance, with all elements of Type 3, plus a sharing of long-term strategies. This includes early supplier involvement in new product/service/process idea inception, concurrent engineering, and the mutual recognition of a long-term, ongoing relationship between the buying and selling firm. Supplier makes important contributions to the firm's competitive success/strategy. The contributions made by the supplier ally are not readily available from other suppliers. They involve ongoing, joint teams, top management support, contact, and visibility.

Source: L.M. Ellram, "Defining Strategic Alliances: Life Cycle Patterns," *Logistics, Navigating the Future,* 78th Annual International Purchasing Conference Proceedings, Tempe, AZ: NAPM, 1993, pp. 59-64.

Expanded partnering extends this philosophy. This type of relationship is very rare in practice. Expanded partnering involves a long-term relationship in which the partners work very closely together for the mutual benefit of both organizations—sharing ideas and perhaps personnel and technology, often through a team. In this type of relationship, the parties recognize their mutual dependence and share strategies as well as risks. This is similar to the idea of an "American Keiretsu," or a strategic alliance on the continuum shown in Figure 5-3. It uses a multilevel network of suppliers, in which first-tier suppliers work closely with second-tier suppliers and so on.

FIGURE 5-4
SEMATECH's Partnering Philosophy

It starts with:
- An attitude and behavioral change at the top of the organization
- Recognition of long-term mutual dependencies internal and external to the organization
- A commitment to this change being understood and valued at all levels within the organization

At the core or basic level, partnering:
- Fosters excellence throughout the organization
- Encourages open communication in a beneficial, supportive, and nonadversarial environment of mutual trust and respect
- Carries this positive environment outward from the organization to its customers and suppliers

At an expanded level, partnering involves:
- Teaming
- Sharing resources
- Melding of customer and supplier
- Eliminating the we/they approach to conducting business

Partnering is not:
- A negotiation or purchasing tool to be used as a lever against the supplier
- A business guarantee

In all cases, partnering promotes:
- A desire and a commitment to excellence through continuous improvements in communication skills, quality, delivery, administration, and service performance
- The factors that contribute to customer satisfaction and the lowest Total-Cost-of-Ownership

---

Source: *Partnering for Total Quality,* Volume 2, © 1990, SEMATECH, Inc. Used with permission.

An Alternative Partnership Classification

Because studies of supplier partnering have shown that one of the greatest sources of confusion lies in understanding what is meant by the term "partnering," Ellram (1993) developed a four-way classification of partnering. This is shown in Figure 5-3. The same system is cross-classified, based on the nature of the desired benefit and the nature of the supplier commitment required, in Figure 5-5.

As SEMATECH and many writers and researchers have suggested, the type of "partnering" relationship must fit with the situation and the

FIGURE 5-5

Partnering Relationship Types Classified by Desired Benefit and Nature of Supplier Commitment

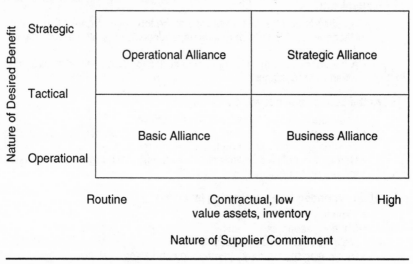

desired benefits. The greater the benefits desired and the greater the degree of supplier cooperation needed, the closer the relationship must be to the "strategic alliance" end of the continuum. Each of the alliance classifications suggested in Figures 5-3 and 5-5 is discussed briefly in the following section. Keep in mind that alliance types are really continuous in nature, rather than discrete. These classifications have been developed for ease of discussion and understanding.

A basic alliance is compatible with SEMATECH's basic partnering. However, the emphasis here is on external relations. In a basic alliance, the organizations treat each other with mutual respect and honesty. Each provides the other with the information required on a timely basis, so that each can perform effectively. This type of arrangement does not involve much, if any, sharing of technical or sensitive data. A company that enters into a long-term agreement with its spare parts supplier to stock the spares needed and provide 24-hour turnaround might fit into this category. There is sharing of information regarding projected needs, pricing, obsolescence, and so on, yet there is no benefit in sharing more sensitive data.

An operational alliance with a supplier includes all the elements of a basic alliance and more. Here, the nature of the desired benefit is important to the buying organization. However, the supplier performs its service or supplies its product as a matter of routine, so it does not represent an additional commitment for that supplier. There may be some joint problem solving for specific issues, but there are no ongoing, cross-organizational teams. An example of this type of relationship might involve an organization that buys a certain type of computer component that is engineered into its product. The supplier makes the component for many customers, and this particular buying organization is not an unusually large customer. In this case, the buying firm wishes to maintain an excellent working relationship with the supplier. The relationship is not so important to the supplier on an ongoing basis that it desires to work closely with the supplier on continuous improvement. Yet the buying organization needs accurate information from the supplier regarding potential component changes, shortages, and prices, because this component is important to its operation. Another example of this might be the relationship of a large, sophisticated organization such as American Hospital Supply with a relatively small hospital.

In a business alliance, the buying organization wants the supplier to provide some unique or specialized product or service. This requires the supplier to invest in additional assets, specialized personnel, technology, or to make some similar type of commitment. While a value-added benefit is received by the buying organization, the benefit is not strategic or core to that organization's business success. Nor does it affect perceived value to the ultimate customer. Because of the uniqueness of the supplier's product or service offering, there is a greater mutual commitment needed for this type of relationship than is needed for an operational alliance. Organizations may use ad hoc, interorganizational teams for joint development.

An example that could be classified as a "business alliance" is a relationship between Thums Long Beach Company and one of its service providers. Thums, a wholly-owned subsidiary of Arco Oil and Gas, is constantly looking for innovative ways to increase oil well production and improve well performance. One of the company's service providers approached Thums with some ideas for improving well performance. The supplier offered to "loan" Arco one of its engineers on a full-time basis to look for low-production wells. That engineer would report to Thums each day, and would look for the wells most likely to benefit from the supplier's well-stimulation technology. The supplier was taking a risk by assigning a full-time employee with no guarantee of additional business. Thums had little to lose, unless the supplier's engineer misidentified an opportunity.

The fourth type of alliance is the strategic alliance. This is rare in practice, because it requires such intense effort and commitment to be successful. Yet the benefits can be substantial. In a strategic alliance, the good or service being provided is of strategic importance to the success of the organization. In addition, to maximize the value-added benefit, the supplier is required to make a special investment. A strong mutual benefit must exist for this type of relationship to succeed.

One example of a strategic alliance is the relationship that Bose Corporation maintains with some of its key suppliers. Bose calls this relationship "JIT II." The steps in developing this type of relationship are shown in Figure 5-6. In JIT II, the supplier actually places one of its employees, called an "in-plant," in the buying company's office, replacing a purchaser, planner, and salesperson. The concurrent engineering and continuous improvement aspects of JIT II distinguish it from what could otherwise be a business alliance, or an operational alliance. One of the companies with which Bose has established this "in-plant" relationship is G & F Industries, an injection molder. The in-plant places orders, practices concurrent engineering, and has full access to all of Bose's facilities, information, and employees. The supplier benefits from greater integration with the customer, improved communications, more efficient administrative processes, and savings on "sales effort," to name a few of the benefits.[3]

Some of the many benefits of alliances have been mentioned in the preceding section. The next section discusses both the benefits and risks of alliances, as well as how to manage those risks.

Benefits and Risks of Alliance Relationships

Establishing and maintaining alliances requires a great deal of time and effort. In addition, such relationships may require a change in corporate philosophy toward suppliers and the commitment of assets. Thus such relationships must yield extraordinary benefits. The nature and extent of the benefits desired and achieved through alliance efforts will vary, depending on the type of alliance relationship. A greater breadth and depth of benefits should accrue as a strategic alliance is approached. However, a strategic alliance also requires the greatest effort and entails the greatest risk. A listing of some of the potential benefits of alliances is shown in Figure 5-7.

---

[3] Porter, Ann Millen. " 'JIT II' is Here." *Purchasing,* Sept. 12, 1991.

FIGURE 5-6
Steps in JIT II Implementation

Steps 1, 2: Supplier reassigns its sales representative to new duties, and customer reassigns its purchaser.

Step 3: In full JIT II implementation, the customer also reassigns its material planner to new duties.

Step 4: Supplier replaces purchaser, planner, and salesperson with a full-time professional person at the customer's location. At BOSE Corp., supplier professionals are called "in-plants." Although supplier replaces purchaser and an in-plant rep, this step actually assists existing purchasing personnel, as more people address the overall department workload.

Step 5: The in-plant representative works 40 hours per week at the customer's location, usually residing in the purchasing department.

Step 6: Customer empowers the in-plant within its planning and purchasing systems. The in- plant works directly from the customer's MRP (or similar) system, and uses the customer's purchase orders to place material orders on his or her own company. Note: Customer typically prohibits the in-plant from placing purchase orders with other companies.

Step 7: Customer provides the in-plant with an employee badge (or equivalent), providing free access to customer engineering and manufacturing personnel. When not planning and ordering material, the in-plant practices concurrent engineering by working with the customer's design engineering staff.

Step 8: Customer and supplier understand that many more steps lie ahead. JIT II will cause change in both organizations.

Source: *Purchasing,* May 6, 1993, p. 17.

Some of the potential risks of alliance relationships are listed in Figure 5-8. These risks must be weighed against the potential benefits of partnering and managed when entering into an alliance relationship. For example, the risk of information leakage can be mitigated by requiring proprietary information/non- disclosure agreements, and sharing information slowly, as trusts builds. The risk of becoming committed to the "wrong" technology can be reduced by having relationships with several suppliers who are developing different technologies. If the organization has a technology or trade secret that, if revealed, would destroy its competitive advantage, the implications must be seriously weighed before sharing it with any outside party, no matter how secure the alliance may appear. The benefits of the proposed alliance must outweigh the risks.

FIGURE 5-7

Potential Advantages of Forming Purchasing Partnerships Versus Adversarial Relationships

MANAGEMENT
- Win-Win approach to problem resolution (shared destiny)
- Reduced supplier base allows closer management/understanding
- Increased mutual dependence lowers risk of losing supply source and creates greater stability through increased supplier loyalty
- Reduced time spent looking for new suppliers/gathering competitive bids
- Loyalty may increase supplier attention and customer service in areas such as:
    - Lead time reliability
    - Priority in times of scarcity
    - Increased attention when problems arise
- Allows for joint planning and information sharing based on mutual trust and benefit
- Greater cooperation from suppliers to support the firm's strategy
- Focus in same strategic direction
- Frequent interaction and feedback promotes understanding of processes, goals, expectations

TECHNOLOGY
- Partners may be more willing to share/give access to technology
- Partners may be more capable of participating in product design based on knowledge and commitment to the other partner
- Supplier knowledge/involvement in design may:
    - Improve quality
    - Reduce time to market for new products/design changes
    - Allow planning for new technologies
- Teamwork to help solve technical problems

FINANCIAL
- May share business risks and rewards through:
    - Joint investment
    - Joint research and development
    - Sharing of financial risks associated with market shifts
    - Joint cost savings programs
- Information sharing/forecasting may reduce inventory levels
- Long-term commitment of a partnership may lead to more stable supply prices
- Focus on total cost of ownership
- Reduced operating costs

QUALITY
- Continuous improvement focus
- Quality training, documentation, and reporting
- Quality process focus

FIGURE 5-8
Potential Disadvantages of Forming Purchasing Partnerships

Loss of competitive drive by supplier
Limit your options/lose flexibility
    Choose "wrong" technology
    Choose "wrong" supplier
Technology leaks
Supplier compromises security or leaks your strategy
Tarnished corporate image if associated with a problematic supplier
High costs to switch to another supplier or terminate relationship
Loss of alternative supply sources

Alliance/Partnership Development

The issues that are important in developing an alliance relationship with a supplier are different than those in a routine supplier relationship. This section discusses the selection, development, maintenance, and dissolution of supplier alliances.

Alliance Selection

In selecting a potential supplier alliance partner, different selection factors may become critical. While price, quality, delivery, and other "normal" factors are still important, other issues that could affect the continuity of the relationship may take center stage. These issues include long term financial stability, management compatibility, willingness to work closely with your personnel, long-term plans, and so on. Many times, a new supplier will not be selected for an alliance relationship. Rather, the organization will use an existing, long-term supplier with which the company has already established a close, trusting relationship. In fact, a recent study of supplier alliances sponsored by the Center for Advanced Purchasing Studies (CAPS) indicated that, on average, firms had a relationship with a supplier for nearly 10 years before the supplier became part of an alliance.[4]

---

[4] Hendrick, T.E. and L.M. Ellram. *Strategic Supplier Partnerships: An International Study.* Tempe, AZ: Center for Advanced Purchasing Studies, 1993.

Alliance Development

As suggested by the CAPS study findings, alliances tend to develop over time, and are rarely implemented overnight. The nature of an alliance relationship will evolve. As trust builds, more information is shared by both parties. A greater comfort level builds, as does mutual dependence. One would expect a purchasing organization to give a much greater share (if not all) of its business to an alliance partner over time.

Whether an alliance is being established with a long-time supplier or a new supplier, it is absolutely critical to put the expectations and parameters of the alliance in writing. This does not necessarily have to involve a contractual commitment. However, as indicated earlier in this chapter, alliances exist on a continuum. Without clarifying the expectations upfront, it is unlikely that both parties will share the same understanding of what the alliance entails. As suggested by a noted expert in the area of interorganizational alliances, a firm should "Establish expectations in writing, then put them away."[5]

Organizations that are considering alliance relationships with new suppliers, or with relatively unfamiliar suppliers, can learn some lessons from those whose alliance relationships have evolved over time. First, if time permits, allow the relationship to evolve. Trust is earned, based upon experience in dealing with one another. Do not take risks by sharing important information early in the relationship. Keep in mind at all times what the organization has to lose, as well as what the firm has to gain. As the organizations demonstrate their compatibility, the relationship can expand, incorporating greater depth and breadth of information, products or services, and volume.

Alliance Maintenance

Once the alliance has become a viable relationship that both organizations would like to expand and build upon, maintaining the relationship requires effort. As in the Bose example, this could involve co-locating a supplier representative within the buying organization as an ongoing part of the organization's team, including concurrent engineering. Other methods of maintaining the alliance include joint continuous improvement efforts, the previously discussed supplier development efforts, and perhaps even purchaser development by the supplier organization. The key idea is to work hard at keeping the relationship from becoming complacent or stale.

---

[5] Ohmae, Kenichi. "The Global Logic of Strategic Alliances." *Harvard Business Review,* Volume 67, No. 2, pp. 143-54.

FIGURE 5-9

Mean Rank Scores of the Action Taken if Received/Provided Consistently Poor Quality and Performance from/to this Supplier/Purchaser

(1 = highest choice)*	Purchasers' Response	Suppliers' Response
Purchaser/supplier work together to identify/solve problem	1.24	1.58
Call in supplier's top management (customer's top management would call us)	1.89	1.96
Reduction of dollar volume	3.35	2.86
Refuse to purchase new items	3.55	3.31
Stop all purchases	4.89	4.44

*Note: There is a perfect rank-order correlation among these five responses.

Source: T. Hendrick and L. Ellram, *Strategic Supplier Partnering: An International Study*, Tempe, AZ: Center for Advanced Purchasing Studies, 1993.

Alliances are developed for the mutual benefit of those involved, which means making the participants more competitive. If complacency sets in, the relationship will be less desirable than open competition in the free market.

Alliance Dissolution

Dissolving an alliance is the last resort for both the buying and supplying firms, as supported by Figure 5-9. Purchasers and suppliers would rather work together as positively as possible, gradually accelerating the consequences of poor performance. However, alliance dissolution may be necessary for two major reasons: first, a business reason no longer exists to maintain the alliance; and second, one or both parties are not performing satisfactorily after efforts have been made to improve performance. An alliance dissolution may be needed if the buying organization discontinues a product or service that supported the alliance, if the supplier exits from the business that supported the alliance, or if some similar reason exists. Organizations frequently change strategic direction. Based on the good relationship established between the alliance partners, the party who wishes to end the relationship should notify the other as soon as possible, so the transition can go smoothly.

On the other hand, if a relationship is being terminated due to the poor performance of one or both parties, the situation may be unpleasant, and a great deal of avoidance and animosity may occur. It is precisely because of these unfortunate situations that each party's responsibility should be established early in the relationship. If there are shared assets, an agreement in writing should indicate the ownership and division of the assets.

Supply Base Reduction/Rationalization

Supply base reduction, or rationalization, has received a great deal of attention both in the press and in practice in recent years. By their very nature, alliances require a reduction in the number of suppliers. Without a reduction, it is not possible to have good two-way communications, create good working relationships, foster continuous improvement, and develop a mutual commitment to one another. This section discusses the benefits and potential hazards of reducing the organization's supply base from the standpoint of all supplier relationships.

Concentrating an organization's business among one or a few suppliers has many advantages. It can result in reduced transportation cost, lower prices, and better service as the organization becomes more important to the supplier. Quality is more readily managed and improves when the number of sources is smaller. Purchasing's administrative expenses are lower, as it manages fewer sources. Purchasing can also use the extra administrative time to work toward developing better relationships with suppliers, supplier development, and ESI.

A single source may be the only economically viable alternative if special tooling or expensive investments are required for a supplier to meet the organization's needs. There may be only one source available due to patents or low industry demand. Many organizations try to avoid single sources whenever possible, due to the risk of dependency and potential supply interruption. Instead, they try to focus their business on two or three suppliers. This allows the organization to get many of the benefits of a reduced supplier base, while keeping a firm hold in the competitive marketplace in terms of price and product/service offerings. In addition to avoiding excessive dependence on the supplier, the buying organization may want to limit the percentage of the supplier's business it accounts for, so the supplier does not become too dependent upon the company. This is particularly true if a buying organization is in a volatile or cyclical industry.

Supply Chain Management

Supply chain management is an approach to managing the flow of information and inventory throughout the entire channel or supply chain, from the earliest supplier to the ultimate consumer—up to and including disposal issues. A systems approach is used to view the activities in the entire supply chain, so information is shared as needed in order to optimize supply chain performance. Supply chain management tends to be cooperative, focusing resources and activities where the process is most efficient and effective, in order to meet the supply chain's customer service goals.

One goal of supply chain management is cycle time reduction. In order to best accomplish this goal, process mapping is often utilized. Hewlett-Packard (HP) uses this approach, modeling the supply chain by looking at the activities that occur at each node or location. HP then analyzes each activity based on the concept that materials arrive at a node, value is somehow added, and the materials leave that node. It is the uncertainty regarding the lead times (supplier performance) and the processing times (manufacturing performance) at each node that creates inventory buildup.

In addition, there may be non-value-added activities or activities that are being performed at the wrong location in the supply chain. These factors, coupled with uncertainty in customer demand, create supply chain inefficiencies. By looking at supply chains as systems, HP has been able to reduce variability, improving supply chain efficiency. An example of this is illustrated by HP's management of the DeskJet Printer, a product that is distributed globally. By postponing the localization of printers—that is, by adding the power cord, transformer, and manual to meet local voltage and language requirements—distribution centers were able to carry lower inventory in the form of one "generic" model. That change reduced the amount of inventory required to meet customer service goals from seven weeks to five weeks. The savings from inventory reduction more than offset the cost of implementing the change.[6]

Another key advantage of supply chain management is the reduction of risk. By sharing assets, information, and overall plans for the supply chain, members of the chain can plan activities and investments with more certainty. For example, GE Medical Systems implemented an integrated supplier program on low-value parts, packaging materials, and MRO items

---

[6] Davis, Tom. "Effective Supply Chain Management." *Sloan Management Review,* Volume 34, No. 4 (1993), pp. 35-46.

because it was not adding value to the supply chain by using its previous approach of ordering from 500 suppliers. By using only six suppliers that coordinate all orders, manage inventory, and check stock, GE Medical Systems has been able to reduce its total cost of ownership while improving service levels. GE Medical Systems looked at the entire system in terms of its *process* related to low value orders, and was able to improve service and cost by reducing supplier uncertainty, while increasing supplier volume.[7] Overall, supply chain management is a growing trend that has the potential to revolutionize the way organizations operate.

Issues in Reciprocity

Reciprocity exists when a purchaser gives preference to suppliers that are also customers of the organization. This is a common business practice today, and many think it makes good business sense to further cement close purchaser-supplier relationships. However, there are times when reciprocal arrangements become illegal. The legal issues associated with reciprocity, the impact of reciprocity on supplier relationships, and international reciprocal practices are discussed in this section.

Legality
It is perfectly legal for an organization to buy from a supplier that is also a customer. As mentioned above, it can be construed as an act of goodwill that can further develop an already good working relationship. Some organizations have a policy of buying from a good customer "if all else is equal." In reality, all else is seldom equal.

Reciprocity becomes illegal if one organization requires another to buy from it, threatening to quit buying from the other if purchases are not reciprocated. This threat may involve purchasing from the supplier/customer at above-market prices, or purchasing a product or service that is somehow inferior to what the organization would buy in an entirely free market situation.

Even when reciprocity is legal, some organizations try to avoid giving any preference to a supplier that is also a good customer, for a number of reasons. First, such a situation may not really be competitive, or purchasing may feel pressured to buy from the customer. Preference could create

---

[7] Ellram, Lisa M. "The 'What' of Supply Chain Management." *NAPM Insights*, March 1994, pp. 26-27.

an antitrust situation due to restraint of trade, and it could hamper the efforts of new suppliers to be seriously considered by the organization.

Impact on Supplier Relations

Reciprocity can have both a positive and a negative impact on supplier relationships, depending upon the angle from which it is approached. For the good customer who receives the buying organization's business as a supplier, reciprocal buying can help strengthen the relationship. However, it may also create complacency and diminish competition.

For sellers who are not currently customers of the buying organization, reciprocal arrangements can hurt the organization's reputation. Requests for bid or proposal may not be taken seriously. The organization could miss out on opportunities for b᠁er prices, quality, or technology, because other suppliers will not want to waste their time with the organization.

International Practices

Reciprocity is common practice with international governments, and it is not illegal. Purchasers who conduct business internationally should be aware of this. It may be impossible to enter a market as a supplier without engaging in reciprocal arrangements through countertrade. Simply stated, countertrade requires an organization that sells and/or produces a good or service in a country to buy goods and/or services from that country.

In some cases, reciprocity is institutionalized within a country, making it difficult to enter that country's markets. One example occurs among the Japanese *zaibatsu* or *keiretsu*. The supply or manufacturing *keiretsu* are characterized by common ownership and interlocking directorates. Not only do the manufacturing firms up the supply chain buy from the suppliers in which they own an equity interest, but these suppliers may be restricted from selling anything outside of the *keiretsu*.

Associations

Another important role the purchasing function plays in representing the organization is through participation in associations. There are a wide variety of associations to which members of the purchasing function may belong. The three major types are trade associations, professional associations, and group purchasing associations. Each is discussed in the following section.

Trade Associations

Trade associations are centered around a particular industry. For example, in the insurance industry, there is the American Insurance Association. In the electrical manufacturing industry, there is the National Electric Manufacturers Association, and in the real estate industry, there is the National Association of Realtors. These associations represent the interest of the industry, rather than representing one profession within that industry. They are often a source of information on trade shows, and they may inform members and lobby regarding legislation that affects the industry.

Organizations and individuals join these associations primarily to keep in touch with current trends in the industry, and perhaps to participate in trade shows sponsored by the association. Purchasing may represent the organization by participating in the association's board, at their meetings on a local or national level, or by representing the organization at trade fairs.

Professional Associations

Professional associations are formed to advance the recognition, practice, and conduct of a profession. The purpose of these associations may include any or all of the following: to educate members; to provide seminars and professional publications; to provide a forum for networking with other persons in the profession; to certify members' knowledge and/or skill levels; to keep abreast of current legislation affecting the trade, which may include lobbying efforts; and to generally advance the recognition of the profession. A number of purchasing associations have emerged as the purchasing function has become more widely recognized and accepted as a profession.

The largest professional purchasing association in the United States is the National Association of Purchasing Management (NAPM), with approximately 35,000 members and over 170 local chapters. In 1974, NAPM instituted the Certified Purchasing Management exam (C.P.M.), which tests the knowledge level of purchasing professionals. Upon successful completion of this four-part exam, coupled with the completion of certain educational and experience requirements, the candidate is designated as a Certified Purchasing Manager (C.P.M.). Recertification is required every five years to assure that continuing education and professional development is occurring. Other professional purchasing associations throughout the world have expressed an interest in the C.P.M. exam and designation. Many are currently considering the adoption of a similar program.

The major professional purchasing association in Canada is the Purchasing Management Association of Canada (PMAC), which was formed in 1919. Around 7,000 members are organized into 53 districts. Like NAPM, PMAC focuses on the education and professional development of members. It has an accreditation program, which was instituted in 1963. This program requires successful completion of seminars, courses,

FIGURE 5-10
International Federation of Purchasing and Materials Management

Argentina	Hungary	Philippines
Austria	India	Portugal
Australia	Ireland	Republic of South
Brazil	Israel	Africa
Costa Rica	Italy	Spain
Denmark	Japan	Sri Lanka
Finland	Malaysia	Sweden
France	Namibia	Switzerland
Germany	Netherlands	Tunisia
Great Britain	New Zealand	USA
Greece	Nigeria	
Hong Kong	Norway	

and an oral examination, upon which the Certified Professional Purchaser (C.P.P.) designation is awarded.[8]

The major professional purchasing associations in many countries, including the Australian Institute of Purchasing and Materials, Ltd., the chartered Institute of Purchasing and Supply in Great Britain, the Japan Materials Management Association, and the Swedish National Association of Purchasing and Logistics, are loosely organized into the International Federation of Purchasing and Materials Management (IFPMM). A list of the country affiliations of the major member associations is shown in Figure 5-10. The IFPMM meets annually, and its primary objectives involve facilitating cooperation, education, and research in purchasing on a worldwide level among the member associations, which currently number over 40.

Other types of professional purchasing associations found in the United States are listed in Figure 5-11. While some of these are focused on the purchasing profession in general, most focus on a specific trade or industry within the purchasing profession, such as the National Institute of Government Purchasing and the National Association of Hospital Purchasing Management. In addition, this figure lists some associations that are not specific to purchasing, but focus on purchasing-associated issues, including the American Production and Inventory Control Society, the Contract Management Association, and the Council of Logistics Management. It is not

---

[8] Leenders, Michiel and Harold E. Fearon. *Purchasing and Materials Management,* 10th edition, Homewood, IL: Irwin, 1993.

FIGURE 5-11
Professional Purchasing Associations

American Purchasing Society
11910 Oak Trail Way
Port Richey, FL 34668-1037
Tel: (813) 862-7998
Hist. Note: APS members include purchasing agents, purchasers, procurement specialists, purchasing managers, purchasing executives and others who buy goods and services. Certifies purchasers and purchasing managers. APS is concerned with improving purchasing performance in business through the education of its membership and the development of ethical standards of conduct in the marketplace. Conducts third-party supplier evaluations.

American Society for Hospital Materials Management
840 N. Lake Shore Drive
Chicago, IL 60611
Tel: (312) 280-6155 *Fax:* (312) 280-4152
Hist. Note: Affiliated with the American Hospital Association.

Coalition for Government Procurement
1990 M Street N.W., Suite 400
Washington, DC 20036
Tel: (202) 331-0975 *Fax:* (202) 659-5754
Hist. Note: Members are firms who provide commercial goods to the federal government, and related associations. Formerly (1988) the Coalition for Common Sense in Government.

Nat'l Ass'n of Black Procurement Professionals
2213 M St. N.W., Suite 300
Washington, DC 20037
Tel: (202) 223-1273

National Association of Educational Purchasers
450 Wireless Blvd.
Hauppauge, NY 11788
Tel: (516)273-2600 *Fax:* (516) 273-2305
Hist. Note: Founded as the Education Purchasers Ass'n, it assumed its present name in 1947. Members are college and university purchasing directors.

National Association of Purchasing Management
2055 East Centennial Circle
P.O. Box 22160
Tempe, AZ 85285-2160
Tel: (602) 752-6276 *Fax:* (602) 752-7890
Hist. Note: Established as the National

Association of Purchasing Agents, it assumed its present name in 1968. Members include purchasing agents, purchasers, procurement specialists, purchasing managers, purchasing executives and others who buy goods and services. Sponsors the Certified Purchasing Manager (C.P.M.) program of professional competency.

National Association of State Purchasing Officials
P.O. Box 11910
Iron Works Pike
Lexington, KY 40578-1910
Tel: (606) 231-1906 *Fax:* (606) 231-1928
Hist. Note: NASPO was established to address purchasing of goods and services by state governments.

National Contract Management Association
1912 Woodford Rd.
Vienna, VA 22182-3728
Tel: (703) 448-9231 *Fax:* (703) 448-0939
Hist. Note: Members are concerned with various forms of contracting with federal, state, and local governments and industry.

National Institute of Governmental Purchasing
11800 Sunrise Valley Drive
Suite 1050
Reston, VA 22091-5302
Tel: (703) 715-9400 *Fax:* (703) 715-9897

National Purchasing Institute
7910 Woodmont Avenue, Suite 1430
Bethesda, MD 20814-3015
Tel: (301) 951-0108 *Fax:* (301) 913-0001
Hist. Note: Members are educational, government, and institutional purchasing administrators.

Printing Brokerage Buyers Association
5310 N. W. 33rd Ave, Suite 114
Ft. Lauderdale, FL 33308
Tel: (305) 735-4111 *Fax:* (305) 733-7161
Hist. Note: PBBA promotes business relationships between brokers, manufacturers and related companies in the printing industry; sets standards and codes of ethical conduct; and acts as a source of information and referral.

Procurement Round Table
4410 Massachusetts Ave. N.W.
Washington, DC 20016-8000
Tel: (302) 885-1555 *Fax:* (202) 885-2966
Hist. Note: Members provide advice on federal procurement policy matters.

Source: *National Trade and Professional Associations of the United States,* 29th edition, 1994. Columbia Books Inc., 1212 New York Ave., N.W., Suite 330, Washington, DC 20005. Used with permission.

meant to be a complete listing. Readers interested in finding out more about these professional associations and others are referred to *National Trade and Professional Associations of the United States,* published by Columbia Books, Inc., or *The Encyclopedia of Associations,* published by Gale Research, Inc. The latter has a version for the United States, an international edition, and an edition that focuses on regional, state, and local organizations.

Through membership in professional associations, purchasers represent themselves as a profession to the community and the outside world. They also represent their employers. These associations often strive to help their membership become more professional, widely recognized, and accepted. Such membership may represent a good forum for educating others about the opportunities and issues associated with the purchasing profession.

Group Purchasing Associations

Group purchasing associations are very different from trade and professional associations. Group purchasing associations have a very specific economic goal in mind: to pool the purchasing power of two or more institutions, in order to gain buying clout and reduce the administrative burden of purchasing. The practice is also known as "cooperative" buying, and these associations often set up group contracts. Members firms can purchase against the contracts at a substantial cost reduction.

Some examples include the Educational and Institutional Cooperative Service, Inc., which is part of the National Association of Educational Purchasers; and Purchase Connections, which is affiliated with Health Resource Institute of Los Angeles. Participation in such organizations can allow the purchasing function to contribute to the success of the firm by reducing the prices paid for certain items, and by reducing the administrative burden associated with the supplier selection and approval process. However, it should be kept in mind that not all group purchasing associations are the model of efficiency and effectiveness described above. A purchaser should thoroughly investigate an association before joining.

The External Role and Perception of the Purchasing Function

From the discussion throughout this chapter, it should be clear that the purchasing function plays a very important role in the organization's external relationships. Today, more than ever, the boundary spanning role of purchasing, working closely with internal customers, suppliers, and professional associations, creates an opportunity for the purchasing function to add value to the organization.

As we've seen, the way in which organizations operate and compete is changing with the advent of globalization and cooperative arrangements with competitors. Likewise, the way that organizations relate to and manage their supplier base is changing, due to supply base reduction, supplier development, ESI, supplier partnering, and so on. As a result, purchasing truly stands at a crossroads. Members of the purchasing function can take a proactive, leadership role in helping their organizations stay current with many of the new purchasing practices. On the other hand, purchasers can sit back and wait for others, such as engineers and users, to manage the changes in supplier relationships. If the latter occurs, the purchasing profession will have missed perhaps the most tremendous opportunity in this century to take a strategic role, and to advance the contribution of the profession. Unless purchasers add value to their organization's processes, the purchasing function will take a step backward, reverting to the role of administrator and "keeper" of systems. The future of the profession lies in the balance.

KEY POINTS

1. The organization can reap many benefits from cultivating and maintaining good supplier relationships.
2. There are many issues that affect supplier relationships, from paying bills on time to being easy to do business with.
3. Supplier education, supplier development, and early involvement of the supplier in design and redesign can improve supplier performance while reducing the organization's total cost.
4. Continuous improvement concepts and goals should be integrated into supplier relationships.
5. Supplier alliances exist at many levels, and should be tailored to fit the needs and goals of the organization.
6. Supply base rationalization is an important way to lower costs and streamline management of supplier relationships.
7. Supply chain management is an important philosophy for managing the flows of information and inventory from suppliers to the ultimate customer—up to and including disposal issues. It reduces inventory while improving customer service levels.
8. Reciprocity can ease supplier relations, and it may be a necessity when doing business globally. In some cases, however, reciprocity can become illegal.

9. Purchasing represents the organization in many trade and professional associations.

10. Purchasing plays a boundary spanning role as a key interface with outside organizations.

SUGGESTED READINGS

Burt, David N., and Michael Doyle. *The American Keiretsu.* Chicago: Business One Irwin, 1993.

Ellram, Lisa M. "Life Cycle Patterns in Strategic Alliance Development." *Proceedings of the National Association of Purchasing Management 92 nd Annual Conference,* Tempe, AZ: National Association of Purchasing Management, 1993, pp. 59-64.

Ellram, Lisa M. "The Supplier Selection Decision in Strategic Partnerships." *Journal of Purchasing and Materials Management,* Volume 26, No. 4, pp. 8-14.

Hale, Roger L., Ronald Kowal, Donald Carlton and Time Sehnert. *Made in the U.S.A.* Minneapolis: Tennant Co., 1991.

Hendrick, Thomas E. and Lisa M. Ellram. *Strategic Supplier Partnerships: An International Study.* Tempe, AZ: Center for Advanced Purchasing Studies, 1993.

Houlihan, John B. "International Supply Chain Management." *International Journal of Physical Distribution and Logistics Management,* Volume 15, No. 1, pp. 22-38.

Lee, Hau and Corey Billington. "Managing Supply Chain Inventory: Pitfalls and Opportunities." *Sloan Management Review,* Volume 33, No. 3, pp. 65-73.

Leenders, Michiel R. and David L. Blenkhorn. *Reverse Marketing.* New York: The Free Press, 1988.

Leenders Michiel and Harold Fearon. *Purchasing and Materials Management,* Tenth Edition. Homewood: Irwin, 1993.

CHAPTER 6

COMPUTERIZATION AND ITS IMPACT ON PURCHASING

Computerization has drastically changed all aspects of information management for virtually every organization. Because purchasing activities are very information-intensive, purchasing has been profoundly affected by computerization. This chapter discusses several aspects of computerization.

This chapter begins with an overview of the importance and goals of computerization, followed by a discussion of general computer uses and how these applications can affect the purchasing function. The importance of user training is also emphasized. Specific applications of computerization to purchasing are then discussed, including basic applications such as inventory control, and more advanced applications, including electronic data interchange (EDI), barcoding, and artificial intelligence. The chapter closes with a discussion of the future applications of information technology in purchasing. This chapter is followed by two appendices. The first discusses and illustrates the basic types of systems and flows in a computerized purchasing system; the second explains basic computer terminology.

COMPUTER BASICS

It is only since the 1960s that computerization has really become a part of business operations. Computerization represents an excellent opportunity to replace routine, manual effort with greater efficiency, speed, and accuracy. Computers can process, store, and manipulate tremendous amounts of data—more so than human processing systems by an order of magnitude.

When an organization implements a computerized purchasing system, the basic functions performed remain the same: recognition of need; specification of need; determination of source of supply and associated price and terms; purchase order preparation and placement; follow-up; receiving and inspection of goods or services; invoice approval and payment; and record keeping. However, many of these tasks are now automated.

The automation of these tasks is a great benefit to purchasing. Many of the tasks mentioned are routine, operating issues. They are really clerical tasks that, while important, do not require a great deal of skill or expertise. Devoting a great deal of time to such tasks detracts from the purchasing function's ability to focus on more strategic issues like improving supplier relationships, as discussed in Chapter 5. An overview of computerizing the basic purchasing processes is included in Appendix 6-1, following this chapter.

Benefits of Computerization

Computerization in purchasing is very commonplace, and is not something to be feared. While computerization may reduce the number of personnel required in the purchasing function, those eliminated should be low-level clerical personnel. Eliminating the routine operating tasks should actually help improve the purchasing function by:

1. Improving purchasing's image, moving it away from the "clerical," paper-pushing function.
2. Providing faster, more accurate access to historical data regarding pricing, performance, order history, and so on.
3. Improving control by reducing clerical errors, and by using reports automatically generated by the system to alert purchasing of potential problems.
4. Allowing for very efficient, rapid ordering of items with existing contracts.
5. Allowing purchasing to spend more time on "value-added," strategic issues.

Thus it is important for purchasers to have a basic understanding of computer systems and related terminology. The purchasing function should try to be aware of the potential applications of computers in purchasing, and how these applications can improve purchasing's performance.

General Computer Uses

The uses and applications for computers in all types of organizations have grown rapidly. This section briefly discusses some of these uses, and their implications for purchasing.

Processing, Manipulating, and Storing Data
Even if computers were useful for nothing else, the ability of computerized systems to store, process, and retrieve large amounts of data has made a major contribution to businesses. In purchasing, supplier history files can be easily maintained and retrieved, providing better decision making and documentation.

Forecasting and Modeling
Computer systems can be used to efficiently recall historical data and manipulate such data, using models or algorithms to develop forecasts. Purchasing can use such systems to identify needs for purchased materials, based on alternative volume projections. Data can also be manipulated to perform "what if" analysis. For example, a purchaser might want to look at the total cost of a contract over the life of the contract, based on various negotiated prices, terms, and conditions. This can be a useful planning tool, and it can be incorporated into a decision support system, a concept discussed later in this chapter.

Some firms, such as Motorola and Texas Instruments, use computer models to analyze and represent the firm's "total cost-of-ownership" for purchased goods and services. This data is then used as a key input when awarding supplier contracts and allocating volume among suppliers. Firms interested in early supplier involvement (ESI) can also use Computer Aided Design (CAD) to model new products, and can exchange the design information with suppliers. The supplier can then manipulate the model of the new product to develop the best value on the component it will provide. This approach has been used successfully by Tennant Corporation, a manufacturer of industrial sweeping equipment. Tennant specifically looks for this capability in qualifying new suppliers.[1]

Graphics
Computers can provide excellent graphic images. Such a capability can be very useful to the purchasing function in tracking supplier performance trends, or in presenting purchasing data visually at meetings or in reports.

---

[1] Hale, Roger L., Donald D. Carlton, Ronald E. Kowal, and Tim K. Sehnert. "Made in the U.S.A." Minneapolis: Tennant Company, 1991.

Word processing

Computerization has revolutionized the creation, modification, and editing of text. Word processing allows flexibility in presentation and ease of change without redoing the entire document. Word processing capabilities form the foundation of desktop publishing. Desktop publishing uses special software to make documents look more appealing. It often incorporates graphics, tables, and other visual devices to create professional looking documents. Many firms have begun to use desktop publishing rather than typesetting to produce forms, manuals, and other repetitively used documents.

These applications are extremely helpful to purchasing in producing impressive, understandable reports and documents. In addition, these applications make it very easy to modify standard documents to fit slightly different circumstances. This saves on clerical efforts within the purchasing department.

Electronic Mail

Electronic mail, often referred to as E-Mail, allows computer users to transmit written messages back and forth through a software program. This may be strictly internal, or it may allow parties outside of the firm to access the mail system. For example, in MCI's "Supplier Guide," the company invites its suppliers to contact them via E-Mail as an alternative to writing or telephoning. This can be a very efficient, cost-effective method of communication.

An excellent example of the use of E-Mail is the state of Oregon's automated request for proposal (RFP) system. Under this system, the supplier is responsible for responding to open RFPs by accessing the state purchasing division's computers. The state is also providing the necessary hardware and software at over 120 selected sites, in order to allow small, minority and women-owned business enterprises access to the information.

This system has yielded tremendous benefits and savings for the state of Oregon. These savings include: elimination of paper and postage for RFP distribution, saving $60,000; decrease in personnel required to maintain the "paper" system, saving $500,000; price reductions due to greater competition, estimated at over $2.8 million.[2]

---

[2]Sunkel, Jill. "RFP-ASAP!" *NAPM Insights,* July 1993, pp. 40-41.

Spreadsheets

Spreadsheets are one of the most popular applications in computer systems today. An example of a spreadsheet used to weight supplier performance criteria is shown in Figure 6-1. Spreadsheets are computerized work sheets or tables that allow computation and manipulation of data. Spreadsheets contain titles explaining the document, numbers to be analyzed, and formulae to perform the data analysis. By changing the numbers or the formulae, it is very easy to use spreadsheets to perform "what if" analysis.

Database Management

A database management system allows application programs to retrieve required data stored in the computer system. The system must store data in some logical way, showing how different pieces of data are related, in order for retrieval to be efficient. In purchasing, this is a critical issue, due to the large volume of data generated that may require analysis at a later date. For example, a buyer may want to see a history of suppliers with which he/she has placed orders for a particular item in the past six months. The database management system must be able to use the item number to reference the order and "pull up" the pertinent data. If the buyer sees that two suppliers have been used, the buyer may then want the system to provide a transaction history with those suppliers over a given time period for all purchased items. The database management system must have the flexibility to sort data in a variety of ways that are meaningful to the user.

FIGURE 6-1
Sample Spreadsheet Application

A	B	C	D	E	F	G	H	I	J
1	Multi-criteria decision making model								
2					**Criterion**				
3	*Alternative*		*On-time delivery*		*Certification status*		*Service responsiveness*		*Scored values*
4	Supplier 1		60.00		70.00		76.00		70.1
5	Supplier 2		75.00		90.00		93.00		88.1
6	Supplier 3		90.00		100.00		90.00		94.5
7	Criteria Weight		.20		.45		.35		
8									

Note: 1. All criteria are rated on a 0 to 100 scale
2. Criteria weights must sum to 100 percent

Performance Measurement

Database management systems help the firm capture data on performance history. This can be very useful in evaluating suppliers, and in providing the supplier with performance feedback. An example of a group that uses such a system is Honeywell's Air Transport Systems Division. The division rates its suppliers monthly on quality and delivery performance. This allows it to stay in touch with supplier performance issues and proactively address potential problems.

Training

Training is absolutely essential for both general purpose computer systems and purchasing systems. Computer interfaces and processes are different from manual applications. Even today, many people are intimidated by computerized systems, because much of the processing is "invisible" to the outside observer. Further, people often fear that they can do serious harm to the system or data within the system if they make an error. That is generally not the case. Most problems can be resolved relatively easily if someone with the proper training is promptly notified.

Personnel should be trained in the basic uses of the computer system. They should also be trained in the specific applications they use, or with which they interface. While many people can "figure the system out" without training, such an approach tends to be cost-ineffective, and many essential applications and features may be missed. Thus it is important to implement formalized training on systems for all new employees, and for existing employees when new applications or systems are made available. The training can be conducted by either internal or external personnel. Training users in the system creates a basic level of comfort on the part of the users. This will help prevent them from bypassing and overriding the system, which can lead to system inaccuracy, costly errors, and loss of data integrity.

Basic Computer Applications in Purchasing

Many of purchasing's activities are very information-intensive. It is precisely this type of activity that is amenable to computerization. Thus data-intensive activities are some of the first purchasing-related responsibilities to become computerized. These activities include inventory control, receiving, and basic purchasing operations.

Inventory Control

Computers can be useful in many aspects of inventory management. One of the most important inventory applications is keeping a "real time" balance of inventory levels and stock availability. Based on this balance, the computer can automatically generate an order or issue an inventory status update to let purchasing know it should review the inventory level for a possible reorder. Through use of computerized models like economic order quantity (EOQ) and materials requirements planning (MRP), the computer system can recommend an order quantity as well as an order date.

A computer system can help manage inventory by tracking historical demand patterns and average inventory, to help the purchasing function determine whether the item is being properly managed and ordered. It can alert the buyer of slow moving items and potential obsolescence. It can also alert the buyer if there is a problem with inventory pilferage or disappearance.

In order for a computerized inventory system to be effective, it must be used by everyone involved with any type of inventory transaction. If personnel remove stock from inventory without updating the computer system, the system will be inaccurate. When accuracy problems occur, people disbelieve the system and override it, exacerbating the problem. Thus consistent usage and updating of inventory records is critical.

Receiving
Computer systems can be used effectively to keep track of receiving and traffic information. When a shipment arrives, distribution center personnel unload, count, and inspect the shipment, verifying the quantities and items received. The bill of lading can then be entered into the computer, where the computer will match it with the purchase order items and quantities. Purchasing and/or accounts payable will be alerted of any discrepancies, so they can follow up with the supplier.

Ford Motor Company uses a more innovative approach to receiving. When a shipment arrives, distribution center personnel immediately check the items and quantities against the open purchase order via a computer terminal. If there are any discrepancies, the order is refused. If the order matches the corresponding open purchase order precisely, the receiving personnel accept the order on the computer. That acknowledgment automatically recognizes that the order was received complete, so no separate entry of the receiving documents is required. In addition, no follow-up is necessary because the order was received as specified. The payment to the supplier is automatically generated at a later date, paid at the purchase order price.[3]

---

[3] Hammer, Michael and James Champy. *Reengineering the Corporation.* New York: Harper Collins Publishers, 1993.

A computerized inventory system can also be used to track defective parts, including rework costs and status. It can generate bills to the supplier to charge for rework costs. It can also monitor payment receipts for rework expenses and track returns of goods.

Purchasing Process

Computerization can be used to enhance all aspects of the purchasing process. In terms of the purchase cycle itself, computers can dramatically reduce the clerical effort needed at each step, as shown in Figure 6-3. In addition, data management capabilities can be used extensively to create a comprehensive purchasing information system.

Some of the database applications beneficial to purchasing are listed in Figure 6-2. Computerization can also be used to help analyze a wide variety of purchasing data and decisions. This includes tracking supplier performance for certification; analyzing price and cost data; examining life cycle cost information; comparing lease versus buy options; and comparing make versus buy alternatives. Computers can also be used to generate bid requests, monitor the return of bids, and compare various bid options.

Computers also make available relatively complex decision support systems, which can provide buyers with "what if" analysis and a good understanding of the implications of a wide variety of options. As previously discussed, computers can also be used to generate reports for management, suppliers, and internal purchasing department use.

FIGURE 6-2
Database/Data Management Applications in Purchasing

- Maintain a database of qualified sources

- Categorize purchase items into A, B, C categories of importance

- Maintain supplier performance history

- Maintain product specifications

- Manage part numbers and assignment of standard part numbers

- Maintain volume history

- Maintain pricing history

FIGURE 6-3

Steps in Purchasing Cycle: Examples of the Potential Contribution of a Computerized Purchasing System

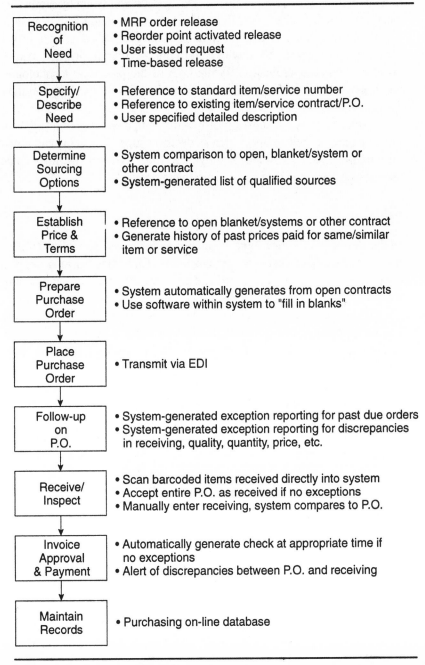

Advanced Computer Applications in Purchasing

The future of computer applications in general, and purchasing applications in particular, is very promising. This section will begin by discussing electronic data interchange (EDI) applications, issues, and implementation. There will also be a brief discussion of barcoding technology, and how that technology can be combined with EDI to improve purchasing and materials management coordination. While both of the preceding technologies are in place in a growing number of firms, there is still much opportunity for growth and improvement.

Next, there is a discussion of several computer systems or applications that are in their infancy, such as decision support systems and artificial intelligence. Artificial intelligence (AI) systems include voice recognition, expert systems, and neural networks.

EDI

EDI is defined as the electronic, computer-to-computer transfer of standard business documents between organizations. EDI transmissions allow a document to be directly processed and acted upon by the receiving organization. Depending on the sophistication of the system, there may be no human intervention at the receiving end. EDI specifically replaces more traditional means of transmitting documents, such as mail and telephone, and may go well beyond simple replacement. EDI has been associated with the term "paperless purchasing," although in most cases, paperless purchasing is more of an ideal than a reality.

There are a couple of key points to note regarding the above definition of EDI. First, the transfer is computer-to-computer. This means that facsimile transmissions do not qualify. Also, the transmission involves standard business documents/forms. Thus E-mail, which is non-standard, free-form data, does not fit the definition. A partial listing of some of the purchasing-associated documents that are currently transmitted via EDI is shown in Figure 6-4.

This section on EDI will discuss the issue of standards, the various types of systems available, EDI benefits and potential problems, EDI implementation, and legal issues.

EDI Standards
In order for EDI to function properly, certain computer language compatibilities are required. First, the users must have common communication

FIGURE 6-4
Partial List of Purchasing-Related EDI Documents

- Purchase Orders

- Material Releases

- Invoices

- Payments (Electronic Funds Transfers)

- Acknowledgments

- Shipping Notices

- Receiving Advises

- Price Lists

- Status Reports

---

standards. This means that documents are transmitted at a certain speed over certain equipment, and the receiver must be able to accept information at that speed from that equipment. In addition, the users must share a common language or message standard. This means that EDI trading partners must have a common definition of words, codes, and symbols, a common format, and a common order of transmission.

The message or language standard is a real issue. There are a multitude of EDI protocols in use today. Some are unique systems created by and for a particular company. Some standards have been adopted within a certain industry. The American National Standards Institute (ANSI) has proposed the use of the ANSI X12 standard. This standard, adapted from the institute's transportation data coordinating committee, is supported by the National Association of Purchasing Management. ANSI X12 supports virtually all purchasing-associated standard documents.

Types of EDI Systems

There are several types of EDI systems in use today, and there are variations on each type of system. The main types of systems are proprietary systems, value-added networks, and industry associations.

Proprietary Systems

Proprietary systems, also known as one-to-many systems, are aptly named, because they involve an EDI system that is owned, managed, and maintained by a single company. That company buys from and is directly connected with a number of suppliers. This situation works best when the company that owns the system is relatively large and powerful, and can readily persuade key suppliers to become part of the network. The advantage for the system owner is control. The disadvantage is that the system may be expensive to establish and maintain internally, and suppliers may not want to be part of the system because it is unique and may require a direct linkage.

Value-Added Networks

Value-added networks, also known as VANs, third-party networks, or many-to-many systems, appear to be the most popular choice for EDI systems today. Under such a system, all EDI transmissions go through a third-party firm, which acts as a central clearinghouse. For example, a buying firm sends a number of purchase orders that go to different suppliers through the VAN. The VAN then sorts the POs by supplier and transmits them to the proper suppliers. The real "value-added" can become apparent when buyers and suppliers use incompatible communication and/or message standards. The VAN can perform translation "invisibly," so the user does not need to worry about system compatibility with its trading partners. This represents a big advantage over one-to-many systems. In addition, the users do not need expertise in EDI standards and issues, as many VANs provide "turnkey," off-the-shelf systems. This can lower start-up costs and reduce start-up lead time.

A value-added network generally utilizes a "mailbox" feature. With this feature, orders and other documents are not automatically transmitted to the receiver when they arrive in the network. Instead, the receiver "picks up" the documents when it so chooses. This may happen at a standard time several times a day, allowing those sending the documents to plan accordingly. This gives the receiver flexibility, particularly if orders are placed or released to be filled at certain times. The user's system is not cluttered with information that will not soon be acted upon.

Yet another advantage of a VAN is that it can be used to receive from, and transmit to, one-to-many systems. This means that the supplier who has customers that use proprietary systems does not need to have a dedicated terminal or direct linkage for each customer. This can increase the ability to network with a customer that uses a proprietary system.

Industry Associations

Industry associations do not represent a completely different form of EDI network; rather, they represent a different *approach* to EDI. Many industry associations have established their own standards for EDI, which are used by member firms within the industry and suppliers to those firms. Examples include the grocery, automotive, retail, warehousing, chemical, and wholesale drug industries.

The advantage of an industry association approach is that it allows compatibility within an entire industry. The VAN can be bypassed, as the firm can use a standard package to do its data translation. This approach can be less costly than using a VAN. On the other hand, suppliers that do not use the industry language will require that the receiving organization use either a VAN or internal translation software for EDI communications. Because some translation is required even within the industry to make transmissions compatible with internal systems, many firms that participate in industry EDI standards still choose to use a VAN.

Incremental Paper Trail

The incremental paper trail is also a variation on EDI. An incremental paper trail is a type of transmission, often through a VAN, in which documents pass through a chain of intermediaries, each of which is somehow involved in the transaction. Each intermediary adds information to the document, rather than creating a completely new and separate document for each transaction. For example, an acknowledgment of receipt could go from the buying firm to the bank, which would authorize payment of the supplier's invoice and the carrier's freight bill. The acknowledgment could then be simultaneously forwarded by the bank to the carrier and the supplier. Both of these parties would then know that the shipment had been received, and that payment was forthcoming.

Benefits of EDI Implementation

It should be obvious from the preceding discussion that EDI is a complex system. However, once in place, EDI tends to be a very easy system with which to interface and communicate. The potential benefits of EDI are many, as shown in Figure 6-5. Most of these benefits are self-explanatory. The reduction in clerical work is a major benefit, reducing paperwork, increasing accuracy and speed, and allowing purchasing to shift its attention to more strategic issues. Cost reduction should also be forthcoming. One expert estimates that EDI can reduce the cost of processing a purchase

FIGURE 6-5
EDI Benefits

- Reduced paperwork to be created and filed

- Improved accuracy due to a reduction in manual processing

- Increased speed of order transmission and of other data

- Reduced clerical/administrative effort in data entry, filing, mailing, and related tasks

- Opportunity for proactive contribution by purchasing, as less time is spent on "clerical tasks"

- Reduced costs of order placement and associated processing and handling

- Improved information availability due to speed of acknowledgments and shipment advises

- Reduced workload and improved accuracy of other departments through linking EDI with other systems, such as barcoding inventory and electronic funds transfers (EFT)

- Reduced inventory due to improved accuracy and reduced order cycle time

---

order by 80 percent.[4] Other firms note that they have been able to reduce their inventory dramatically due to improved inventory accuracy and reduced order cycle time. Despite all of these benefits, there are some drawbacks to EDI implementation. Most of these drawbacks can be overcome, as discussed in the next section.

Drawbacks of EDI Implementation
One of the biggest issues in EDI implementation is the confusion over standards, and uncertainly about the type of system to adopt. There are many standards available, and there is also the choice between one-to-many or VAN systems. The firm must decide whether it should attempt to develop and implement EDI internally, or use an outside party for all or part of the

---

[4] Carbone, James. "Make Way for EDI." *Electronics Purchasing,* September 1992, pp. 20- 24.

implementation effort. Turnkey systems are available. These are standard EDI packages that the firm can install and use relatively cheaply, often interfacing with a VAN.

The whole technological issue is very confusing. This could explain why so many firms are turning to outside consultants to help them with EDI implementation, and are using VANs for data translation. Even very large firms like Coca Cola, Inc. have chosen to use a VAN rather than a one-to-many system. Yet, even if the firm decides to use a VAN and an outside consultant for implementation, there are many options among consultants and VANs. Therefore, it is recommended that firms use a team for EDI implementation. There are no easy answers. Some of the issues that must be addressed are discussed later in the implementation section.

Another drawback of EDI is the cost. There are costs associated with the system set-up and implementation, as well as costs for training users. There are also ongoing costs associated with maintaining the system. If a firm uses a VAN, it incurs monthly user fees and transaction costs. However, most firms feel that EDI more than pays back all of these expenses in increased efficiency and internal transaction cost savings.

Data security has been brought up as a concern with EDI usage, particularly when using a VAN. However, mailbox systems are set up so that each user has access only to its own mailbox. Other techniques, such as authentication and encryption, prevent outside parties from tampering with or falsifying any messages that are sent.

Some have also expressed concern that EDI usage will not really reduce the paperwork, because users will still want a hard copy of everything. This can be a problem. But over time, as the users become more comfortable with the system and with viewing information on-line, this problem should decrease and disappear altogether.

Finally, there is concern about the impact of EDI on supplier relationships. Some fear that EDI will dehumanize supplier relationships, while others fear that it will "lock them in" with a particular supplier. Experience has shown that neither is the case. Rather than dehumanizing relationships, EDI helps buyer-supplier communications by focusing on substantive issues. Further, EDI does not lock in relationships. In a firm using a one-to-many system that provides the supplier with the terminal, change may be delayed. However, because suppliers can still access a customer with a proprietary system using a VAN, fewer organizations are providing suppliers with terminals. Some of the legal questions associated with EDI purchasing are discussed in a separate section.

EDI Implementation

EDI implementation represents a major effort. As such, it should be undertaken with care and planning, actively involving qualified personnel and outside parties as necessary. The following section outlines a sequence of steps that serves as a guideline for EDI implementation.

1. Establish the need for EDI. How would the firm benefit? Would it provide a good payback, or is it just something the organization is interested in because it is a trend?

2. Form a project/planning team to oversee the organization's EDI development and implementation. The team needs a strong leader who supports EDI implementation fully. Members of the team should include purchasing, management information systems, and other affected areas, such as accounting/auditing, legal staff, and manufacturing.

3. Audit the current purchasing system and EDI environment. This includes examining current purchasing practices and the current computerized system, and making necessary changes *before* EDI implementation. In addition, this involves reviewing the firm's external environment to determine which systems competitors, others in the industry, and suppliers are using. If a successful pattern exists, following this pattern should be strongly considered.

4. Determine basic systems specifications. This includes whether the organization should select a VAN or a proprietary system. As discussed above, VANs seem to be the preference today. In addition, the type of language/standard should be determined. Whether the system will be primarily implemented and developed internally or by an outside party should also be considered at this time.

5. Establish an implementation plan. This includes a timetable, recommended documents to implement in order of priority, which suppliers to involve early and which to add later, and so on.

6. Present the plan to top management for approval, if required. Even if approval is not required, this would be a good time to inform top management of the team's decisions and plan.

7. Provide education and training both internally and for potential outside users. This is critical to establish an understanding of the system and to ease implementation. Educational materials should also be developed for employees and suppliers that begin to participate in the system later.

8. Conduct the pilot test of the system with predetermined suppliers. This is often done with sample data, rather than real transactions. It helps to have suppliers participate, particularly those with EDI experience, to assure that interfaces are working properly.

9. Evaluate the system based on the pilot test and a thorough review. Make any changes that are necessary.

10. Develop and expand the system to include additional suppliers, divisions, and documents.

Use of EDI to Support Corporate Strategy

EDI is more than just an operating system to increase the efficiency of the organization's transactions and its exchange of standard business documents. EDI can also play an important role in supporting the strategy of the purchasing function, which in turn will support the organization's strategy.

A key objective of virtually all organizations today is to reduce costs. As discussed, EDI reduces the cost of creating, processing, and managing a variety of transactions. EDI can also reduce costs via inventory and personnel reduction.

EDI can help support the firm's overall materials strategy by providing a linkage with other systems, such as inventory control and receiving. Further, EDI in international applications can dramatically reduce order cycle time from suppliers, again saving money. Perhaps more importantly, EDI can reduce the *buying organization's* cycle time to its end customers by reducing inbound materials lead time. This is true on a local, national, and global scale. Such reduction can help support one of the key objectives of virtually every organization today: faster speed to market.

Finally, the use of EDI can take the focus of the purchasing function away from routine tasks and paperwork. This can allow the purchasing function more time to focus on value-adding activities like partnering, ESI, and supplier rationalization.

Legal Implications of EDI

The jury is still out on the legal status of EDI transactions. Under the Uniform Commercial Code, contracts for sale of goods over $500 must be in writing. In order to get around the issue of "writing," some organizations issue what is known as "trading partner agreements." This is similar to a blanket contract that covers ongoing transactions between parties. It should include authorized ordering parties, terms and conditions, any special authorization code to be included in each order, and terms of dispute resolution. This is simply a safeguard in case there is any disagreement over transactions. Federal and most state court systems will accept hard copies of electronic records as evidence.[5]

---

5 Skupsky, Donald S. "Keeping Records in the Electronic Age." *NAPM Insights,* May 1994, pp. 6-7.

Bar Coding

The topic of bar coding is discussed in some depth in Volume Three of this series. To reiterate the basic concepts of bar coding, a bar code is a sequence of parallel bars of various widths, with varying amounts of space between the bars. The pattern and spacing of the bars convey information such as letters, numbers, and special characters. These bars are optically read by "scanning" them with a beam of light. The information contained in the bars is then read directly into a computer, or is stored and downloaded into the computer system at a later time.

Bar coding can be useful in purchasing. Texas Instruments has linked EDI and bar coding in the order placement and management of office supplies, with positive results. The company reduced the amount of cash tied up in inventory by $2 million, freed up 40,0000 square feet of warehouse space, reassigned 11 office supply control employees, and reduced cycle time by more than one-third.

By bar coding inbound shipments, purchasing and the entire materials function can get more accurate accounts of actual receipts. The bar code error rate has been quoted as between one in ten thousand and one in one million, versus one in 25 or 30 for manually keyed data.[6] Receiving can also be automated, which further contributes to cycle time reduction. This data can automatically be used by the accounts payable department for generating checks and reconciling invoices with purchase orders and receiving. Thus, bar coding represents a logical extension of the organization's information systems, and an excellent linkage with EDI.

Electronic Catalogues

Another computer trend, which is also an excellent tie-in with EDI, is the use of electronic catalogues. Again, these are commonly used with office supplies and other goods that are not used in the direct creation of the organization's final good or service. Such a system is used by Texas Instruments in conjunction with its barcoding/EDI system.

An electronic catalogue system is also used by Digital Equipment Corporation. The supplier provides a tape or similar medium containing the

---

[6] Hatchett, Ed. "Combining EDI with Bar Coding to Automate Procurement." *1992 NAPM*, Tempe, AZ: National Association of Purchasing Management 1992, pp. 45-50.

catalogue. Authorized requisitioners then order the goods from the catalogue, without any intervention or approval by purchasing. The order is transmitted via EDI and drop shipped to the requisitioner. Invoicing is done directly to Digital's accounts payable department via EDI. Payment is made using Electronic Funds Transfer (EFT).[7] EFT is the electronic transmission of funds from one account (Digital's) to another account (the supplier's).

Digital has enjoyed improved service, reduced cycle time, stockroom and inventory reduction, and improved productivity for requisitioners, accounts payable, and purchasing personnel. Such a system allows the purchasing function to be involved only when they can add value to the process. In this case, value can be added in establishing the system, selecting the suppliers to participate, negotiating the prices and items to be included, and establishing terms and conditions. There is no value added by the purchasing function in placing day-to-day orders. Thus purchasing is freed up to concentrate its efforts on areas where it can best contribute to the firm.

Decision Support Systems

Decision support systems (DSS) encompass a wide variety of models and applications that are designed to ease and improve decision making. These systems incorporate information from the organization's database into an analytical framework that represents relationships among data, simulates different operating environments such as price or volume levels, may incorporate uncertainty and "what if" analysis, and uses algorithms and/or heuristics. DSS actually present an analysis and, based upon the analysis, recommend a decision. The artificial intelligence tools discussed in the next section can be incorporated into DSS. DSS may also contain decision analysis frameworks, forecasting models, simulation models, linear programming models, and so on. They can be used to assist in a wide variety of purchasing decisions, such as alternative negotiation options, forward buying and futures trading, and supplier selection under uncertainty. While the use of DSS is not currently widespread, it appears to be growing as its utility becomes more understood and as computing costs continue to decline.

---

[7] Bishop, Nancy, Holger Ericsson, and Sharon Rashid. "Digital's End User Execution Purchasing-Leveraging EDI." *1992 National Association of Purchasing Management Conference Proceedings,* Tempe, AZ: NAPM, 1992, pp. 93-98.

Artificial Intelligence

Although artificial intelligence (AI) has been around since the 1960s, applications of AI to business processes are still in their infancy. Artificial intelligence encompasses a wide range of technologies aimed at making computers reason and/or function like human beings. As such, AI includes areas such as voice recognition (particularly natural language), optical recognition, and expert systems for problem solving.

Natural Language Recognition

Mead Corporation has used the natural language capabilities of AI to make data stored within its computer system much more accessible. At one time, a simple request like the past usage of a commodity could take an hour, involving sorting through pages and pages of reports. Mead's Decision Support Department suggested that the company create an on-line data base that would be voice accessible, using natural language processing. With the advent of this system, the above mentioned data can be accessed in three minutes by verbally asking the system a question. This has contributed tremendously to the productivity of the purchasing function as well as other functions.[8]

Expert Systems

Expert systems are a type of artificial intelligence designed to mirror the problem solving processes of human experts. These knowledge-based systems are generally modeled after an effective, experienced human problem solver. They work best if the problem solving is based on experience and reasoning that cannot be mathematically modeled. They are based on the assumption that the problem solver goes through a number of logical steps, following a set of "rules" in solving the problem. Programming such systems requires that a skilled interviewer ask a human expert the right questions, in the right order, to uncover the rules behind the expert's effective decision making. Such systems can be used to train new employees and to make effective decisions in complex task environments when an expert is not available. Some possible applications for these knowledge-based systems in purchasing are shown in Figure 6-6.

Neural Networks

Neural networks are still in the development stage. They can be considered an offshoot of expert systems, in that they aid decision making by using

---

[8] Bramble, Gary M., Bette Clark, and Robert Florimo. "Artificial Intelligence in Purchasing." *1990 National Association of Purchasing Management Conference Proceedings*, Tempe, AZ: NAPM, 1990, pp. 186-190.

FIGURE 6-6
Potential Expert Systems Applications in Purchasing

- Supplier Selection

- Supplier Qualification

- Supply Base Rationalization

- Analysis of Bids and Proposals

- Determination of Contracts

- Creation of Statement of Work for Service

- Assess Potential Patterns/Problems in Supplier Performance

logic and rules. A key difference is that neural networks have the ability to learn. Neural networks create their own rules based on past decisions and outcomes, rather than relying on an "expert." Once developed, these systems will be excellent for any repetitive activity that requires analysis of more data than a human could effectively process. As such, neural networks could be used to alert management of potential problems in supplier performance patterns, quality, delivery, invoicing, and similar issues.[9]

Impact of Computerization on the Future of Purchasing

Just as computerization has revolutionized business in general, it will continue to have a profound impact on purchasing. Many computer technologies, such as DSS, EDI and bar coding, are currently underimplemented and underutilized. Other technologies, such as expert systems, are still in the development stage, and represent a tremendous opportunity for the future.

One of the challenges for the purchasing function is to be aware of new information technologies and how these technologies can apply to purchasing. Integrating the various technologies currently available is also a major challenge.

---

[9] Yacura, Joseph A. "Supply Line Management Information System." *1992 National Association of Purchasing Management Conference Proceedings,* Tempe, AZ: NAPM, 1992, pp. 343-48.

Historically, when computerization occurs, it tends to focus on automating transactions. Yet there is a much greater opportunity to add value through computerization. The purchasing function is extremely data-intensive, and it can benefit greatly from improved database management systems and decision support systems, and by incorporating features of artificial intelligence into purchasing systems. All of the aforementioned computer enhancements can reduce the purchasing function's clerical workload, support improved decision making and analysis, and allow the purchasing function to better support corporate strategy.

However, the demands on most organization's computing resources are very high. It is up to the purchasing function to be aware of the management information systems projects that are underway, and to understand the potential impact of these projects on purchasing operations. Many organizations are currently undergoing massive changes in their computing systems, so the time to be informed is now.

· In addition, the purchasing function must make its information system needs and wishes known, so that these can become incorporated in the organization's overall management information systems plan and budget. This is another example of an outstanding opportunity for the purchasing function to elevate its status and contribution within the firm. To capitalize on the opportunity, the purchasing function must be proactive and meet the challenge of today's rapidly changing computing environment.

APPENDIX

Basic Overview of Computer System

This appendix provides basic, general information on various types of computer systems and security issues. It also discusses computerization of fundamental purchasing operating tasks, and how computerized information can be used to generate management reports.

Major Types of Systems

Three major types of computer systems are prevalent today: mainframe, mini-computers, and microcomputers. All of these computer systems require both hardware (equipment) and software (programs to run the system and applications). Hardware includes input and output devices, such as terminals and printers, a processor to perform the data manipulation, and data storage devices.

Mainframe systems are powerful, large-scale systems that can manage and manipulate large databases and many software applications simultaneously. These are centralized systems that many people can access through terminals, or by using a microcomputer as a terminal. They may be networked with other systems, such as minicomputers and microcomputers, so that data available on the mainframe can be "downloaded" for use and manipulation on other systems.

Due to the increasing power and capability and the decreasing price of computers, many firms are moving toward the use of minicomputers (small-scale systems). Today's minicomputers have as much power as the mainframe systems of the early 1980s. Through local area networks (LANs), many users can access data, software, and other systems features, just as they would with a mainframe computer. The major difference is that the minicomputers are less powerful and can accommodate fewer users. Just as with mainframe computers, the user can access the system through a terminal or a microcomputer. It is generally transparent to the user, whether he or she is on a mainframe or a minicomputer system.

Microcomputers, also commonly called personal computers (PCs), have revolutionized business in the 1980s and 1990s. These small, stand-alone systems allow individual users to manipulate data, perform word processing, and utilize many other functions independent of the organization's main computer systems. PCs are also frequently networked, or connected, with minicomputer or mainframe computer systems. This gives the PC user access to all of the features of the host system.

Security Issues

Security and controlling access to information have been a concern as long as computer systems have existed. Organizations should institute a general computer control system that limits access to the computer system hardware to authorized personnel. This includes having back-up files of the system on tape, disk, or CD-ROM at a secure location, physically separated from the main computing operations. Applications controls are also needed to prevent access to confidential information. In order to prevent access to data by unauthorized personnel, security systems must be planned and built into the system. One of the most common types of security measures is the use of a password, which is unique for each user and must be changed frequently. The password determines which files the user is allowed to access.

An extra measure used to prevent access to data by an unauthorized user who manages to enter the system is the placement of "passwords" or "keys" within a file. Two popular approaches are encryption and authentication.

These approaches are often used when transferring data externally, such as the sending and receiving of EDI documents. Authentication is like a unique electronic "signature" assuring that the sender is legitimate. Authentication combines the electronic signature with a complete data integrity check of the message. If any of the message is missing or has been altered, the message will not pass the test.

Encryption involves encoding or "scrambling" a file, so the message cannot be read without the proper encryption "key" or password.[10] If an unauthorized user accesses an encrypted file, he or she encounters an unintelligible, scrambled mass of letters, numbers, and symbols.

More advanced, complex methods of data security are constantly being developed for information of a highly confidential, sensitive nature. Discussion of those methods is beyond the scope of this text.

File Structures and Basic Purchasing Information Systems

As illustrated earlier in Figure 6-2, computerization can be used to enhance virtually all aspects of the basic purchasing ordering process and operating system. In order to support the basic purchasing operations, a number of files are required. These are listed and briefly described in Figure 6-7. These files provide access to a great deal of valuable purchasing information in a variety of usable formats.

In addition, data from these files can be used to generate reports that help control internal purchasing operations and that keep management informed of purchasing activities. In terms of operations control, the computerized purchasing system can be set up to include a number of automatic exception reporting features. For example, if an order is past due, the system can automatically generate a report providing information on that order, so purchasing can follow up. If a credit for returned goods takes more than a certain amount of time to be received, a report can be generated to notify purchasing. Such reports eliminate much of the clerical effort and paper shuffling that purchasing would otherwise have to do.

Files can be used to generate management reports that summarize purchase activity. For example, the price files can be used to generate charts and graphs of prices paid and projected prices for a particular good or service. Bid files can be summarized and excerpted efficiently by the computer system to provide management with a summary of bids for a particular purchase. Such

---

[10] Mules, Glen R.J. "EDI Security and Control." *NAPM Insights,* June 1992, pp. 12-13.

FIGURE 6-7
Files Required to Support Basic Purchasing Information System

File	Potential Uses
Supplier Files	Prices, terms, products and services supplied, location, contact persons
Purchase Order/Requisition Files	Information on current and past purchase orders and requisitions
Bid Files	Information on bids solicited, received, and accepted
Price Files	Supplier quotations and actual prices paid referenced by item/service
Follow-up/Expediting Files	Provide status/history of open orders
Delivery/Return Files	Track returns, credits
Acknowledgments	Provide current order status and shipment status
Commodity Files	Provide a history and projection of prices, prior purchases, and demand by item or service
Contracts	Provide documentation of legal agreements, blanket orders, terms and conditions
Total Cost/Life Cycle Cost Analysis Files	Provide estimated and actual cost data associated with the total cost of a good or service

information can also be used strategically to support operations. For example, commodity files can be reviewed and maintained to keep purchasing aware of potential shortage situations or opportunities for forward buying, or for tracking of pricing trends. An awareness of such information can allow purchasing to strategically position the organization, so it can take advantage of changes in price and availability. This gives the company a competitive edge.

Networking
The use of a network to link computers is increasingly common today. NAPM defines a computer network as "a group of physically connected computers (or workstations) that are attached to a file server. The file serv-

er is usually a high-speed computer with abundant memory to control the exchange of data between workstations."[11]

One reason that networks have become popular is that they allow users to share resources. This includes access to data files, so that everyone has the same information, and access to software programs. Networks also allow the sharing of peripheral equipment, such as laser printers. This can save an organization investment dollars and allow for better equipment utilization.

There are two major types of networks. Local area networks (LANs) are connected by cables (hardwired), and are generally confined to one physical location. Wide area networks (WANs) can link a number of computers at locations throughout the world, by using telephone lines.

KEY POINTS

1. There are many benefits to computerization in terms of reducing purchasing's clerical/administrative load, and freeing time for more value-added activities.

2. There are many generalized uses for computers, from data sorting to report preparation.

3. Training on computer systems is vital in order to fully utilize the potential of computers.

4. Basic computer applications in purchasing include inventory control, receiving, and general process automation.

5. There are many advanced computer applications in purchasing, such as EDI, bar coding, decision support systems, and artificial intelligence.

6. Overall, computerization is having a profound impact on the automation of purchasing duties. Computerization is moving from a transaction-oriented focus to a strategic focus.

GLOSSARY OF BASIC COMPUTER TERMINOLOGY[12]

Application Software. A program written for or by a user that is employed to satisfy decision making support or informational requirements associated with a particular use. Examples are WordPerfect, Lotus 1-2-3, Dbase IV, and Nomad.

---

[11] *C.P.M. Study Guide Supplement,* Sixth Edition. Tempe, AZ: NAPM, 1994.

[12] *The C.P.M. Study Guide,* Sixth Edition. Tempe, AZ: NAPM, 1992.

Batch Data Processing. This refers to the processing of data accumulated over a period of time in a single run, usually in a serial fashion. It is usually associated with data entry of a large quantity of related documents.

Central Processing Unit (CPU). This is the centerpiece of the computer system or, strictly speaking, the computer itself. It is composed of the control unit (which decodes program instructions and directs other components of the computer to perform the task specified in the program instructions); the arithmetic-logic unit (which does multiplication, division, subtraction and addition and compares the relative magnitude of two pieces of data); and the primary storage unit (which stores program instructions currently being executed and stores data while it is being processed by the CPU).

Computer (Hardware). An electronic data processor that can perform substantial computation tasks, including numerous arithmetical or logical operations, and execute stored programs. It consists of the CPU, input/output devices, and primary storage devices. A computer has the ability to compare one piece of data with another (logical operations), the capacity for storage and retrieval of data, and the facility to modify (branch) the execution stream of a program based on the values of the input data, all at extremely high speeds.

Computer System. Integrated combination of devices (e.g., input, processing, storage, and output) linked together.

Data. Information or pertinent facts and figures that constitute inputs to the computer.

Database Management System. This is a program that serves as an interface between applications programs and a set of coordinated and integrated physical data files known as database.

Distributed Data Processing. This refers to the distribution of local data processing activities among different computers at an organization's remote locations, thus allowing local data processing needs to be handled by each remote location's own computer.

File. Set of related records treated as a unit (e.g., in-process inventory file).

Information. Results derived from the data that has been processed by the computer.

Input Devices. Input devices include cathode ray tube (CRT) terminals, optical scanners, voice recognition instruments, and readers of magnetically coded tape or disks. Some of these will also operate as output devices.

Integrated Word/Data Processing. This refers to the combining of word processing and numeric data processing capabilities into one integrated system, producing management reports with both text and numbers.

Local Area Networks (LANs). This describes an interconnection of computers, terminals, word processors, facsimiles, and other office machines and/or equipment within a defined area or location.

Management Information System (MIS). This refers to a system for providing information used in decision making.

Output Devices. These include magnetic tapes and disks, printers (impact, print chain, ink jet), laser imaging devices, computer output microfiche (COM) cards, voice output devices, graphics terminals, and CRT terminals. Some of these will also operate as input devices.

Parallel Data Processing. This describes the manipulation of data required or generated by some process while some other process is in operation.

Real Time Data Processing. This describes the manipulation of data as it enters the system so that the information in the system is updated concurrently.

Program (Software). A set of instructions for a particular application (run) or the performance of a specific task that the computer will execute with accuracy and reliability. It is essentially the perceived or actual capability to mimic human decision making and the processing of data or information.

Secondary Storage. This refers to media used for relatively long-term storage of data. The most widely used are magnetic tape and disk. While most data used by the computer is stored in secondary storage, such data must be transferred to primary storage for processing by the CPU. As part of the CPU, the data can be more easily and quickly accessed.

Systems Software. A set of programs that controls the use of the hardware and software resources, and that allocates computer system resources to application programs based on needs or priorities. Examples are MS-DOS, OS/2, and VM.

SUGGESTED READINGS

Carter, Joseph R. and Gary L. Ragatz. "Supplier Bar Codes: Closing the EDI Loop." *International Journal of Purchasing and Materials Management,* Summer 1991, pp. 19-23.

Emmelhainz, Margaret A. "Electronic Data Interchange in Purchasing." *Guide to Purchasing,* Fourth Volume, 1.17, pp. 1-11, 1986.

Leenders, Michiel, and Harold E. Fearon. *Purchasing and Materials Management,* Tenth Edition. Homewood: Irwin, 1993.

Monczka, Robert and Joseph R. Carter. "Implementing Electronic Data Interchange." *Journal of Purchasing and Materials Management,* Volume 25, No. 1 (1989), pp. 26-33.

Sriram, Ven and S. Banerjee. "Electronic Data Interchange: Does Its Adoption Change Purchasing Policies and Procedures?" *International Journal of Purchasing and Materials Management,* Volume 30, No. 1 (1994), pp. 31-40.

CHAPTER 7

SOCIAL RESPONSIBILITY AND THE DISPOSAL OF HAZARDOUS WASTE

This chapter discusses the role of the purchasing function in the purchase, control, storage, and disposal of hazardous and regulated materials within the larger corporate directive of "social responsibility." It begins by examining the role of social responsibility in corporate strategy and its importance to the purchasing function. Hazardous and regulated materials will then be defined, and the legal issues associated with the utilization, production, and disposal of these materials will be discussed. The chapter closes with proactive purchasing strategies related to the minimization, prevention, and control of hazardous and regulated materials.

SOCIAL RESPONSIBILITY

Corporate social responsibility begins with the identification and recognition of the various interests held by the stakeholders of a business entity. Stakeholders are defined as any group that has a vested interest in the operations of the firm. They may include the shareholders, employees, customers, competitors, suppliers, governments, unions, and the local, national, and international community (see Figure 7-1).

R. Jerry Baker, C.P.M., the executive vice president of the National Association of Purchasing Management (NAPM), has indicated that corporations can have a profound impact on their environment, and should therefore "assume all the duties being a good citizen entails."

> Good citizenship on a corporate level is much the same as good citizenship on an individual level. It means supporting the greater community of which we are all a part. But because the good corporate citizenry is

FIGURE 7-1
Firm's Stakeholders

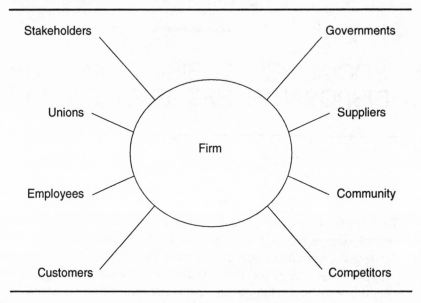

relatively larger and more powerful, it should be ethically obliged to give back to the community on a larger scale than is possible individually. Thus each participant in the corporate citizenry is obliged to do his or her own job in a manner that reflects the corporation's good citizenship. For purchasing, this means taking a proactive role in the corporation's strategic issues.

Ideally, these strategies should work toward supporting public policy in the public interest. Such strategies toward social responsibility can be demonstrated many ways, including equal opportunity employment for minorities, small business set-asides, and environmental accountability, to name a few.

It is purchasing's challenge, then, to not only remain on the sharp edge of profitability, but to find cost-effective means of supporting socially responsible strategies. For instance, purchasing can be instrumental in environmental issues such as waste avoidance and removal, especially when toxic materials are involved.[1]

---

[1] Baker, R. Jerry, C.P.M. "Our Changing World: The Social Responsibility of Purchasing." *NAPM Insights,* February 1993, p. 2.

The challenge to the purchasing function therefore entails going beyond the legal requirements with respect to hazardous and regulated materials by proactively investigating and implementing environmental solutions that recognize the needs of all stakeholders.

Going Green

It seems that almost everything to do with the business world is going, or has gone, green—green marketing, green financing, green computers, and even green purchasing. Going green entails a philosophy of doing business in an environmentally friendly manner. The market-driven world of business is following public opinion and gearing itself toward environmentally friendly practices. Historically, the catalyst behind the green movement has been government regulation, beginning with the Clean Water Act of 1972. In today's marketplace, the pressure is coming from several sources representing "the voice of the stakeholders," including public sentiment, governmental regulations, and even the corporate culture. This new era is resulting in significant business opportunities as well as costs.

The "green business revolution" is clearly gaining momentum as consumers demand green (or environmentally improved) products and services that don't "cost the earth." This trend will continue for three powerful reasons: consumers will demand it; governments will require it; and corporations will find they can use it to strengthen the bottom line.[2]

Tragedies such as Love Canal, New York, in which hundreds of families were displaced because 22,000 tons of hazardous waste was buried in a Niagara Falls subdivision between 1942 and 1953 by the Occidental Petroleum Corporation, have given strength to the "not in my back yard" syndrome. In fact, it is becoming the "not in anyone's back yard" syndrome. Communities are no longer willing to accept monetary compensation for toxic waste disposal sites near their homes. Citizens are unwilling to allow corporations to degrade the natural environment for profit.

> Today's scenario might be described like this: Any industrial facility making anything—automobiles, automotive components, or automotive antifreeze—owes its existence to the community with which it coexists. If the facility fouls the atmosphere and the groundwater sufficiently, the

---

[2] Carson, Patrick, and Julia Moulden. "Green is Gold." *Small Business Reports,* December 1991, pp. 68-69.

community will begin to withdraw, to move away, and to estrange itself—jobs or no jobs.[3]

A recent poll by the consulting firm Arthur D. Little found that degrading the environment is seen as a more significant corporate offense than price fixing or insider trading, as judged by public perception.[4] In addition, 9 out of 10 consumers indicated that they would be willing to pay 6.6 percent more for "green products."[5] Along with this heightened sense of awareness come potential new issues. Hence the need for proactive purchasing strategies to capitalize on them.

Developing a Green Strategy

The "Green Movement" has gone far beyond corporate rhetoric. When the 3M company pioneered its PPP ("Preventing Pollution Pays") environmental message, many corporations ridiculed the concept.[6] Now, less innovative corporations must play catch-up to a strategy that has always made sense but took time to gain acceptance.

Going green requires a commitment, and then it requires a strategy. The strategy should include a statement of basic principles, an outline of the direction the firm will take, identification of the areas the firm will concentrate on, and some sort of time frame.[7]

Across the board, any strategic change within a corporation requires a commitment of corporate resources. Commitment can take many forms—education of employees, capital investment, changes in policies and procedures, and an annual budget. To secure this commitment requires the support of upper management. The majority of the organizations that have made significant strides on environmental issues have had an environmental champion at the highest levels in the corporation, including AT&T, Dow Chemical, First Brands Corporation, Herman Miller, and 3M. The ideal environmental corporate champion should possess a personal value system that embraces environmental concerns, a system that will transfer from the individual's personal life to corporate/public life.

First Brands Corporation, headquartered in Danbury, Connecticut, is a Fortune 300 company that produces many well-known products, including

---

[3] Bergstrom, Robin P. "On Giving Due Thought to Afterthought." *Production*, November 1991, p. 46.

[4] Carson and Moulden, p. 68.

[5] Wasik, John. "Market Is Confusing, but Patience Will Pay Off." *Marketing News*, October 12, 1992, p. 16.

[6] Wilsher, Peter. "The Feeling Grows That Going Green Is Good for Business." *Management Today*, October 1991, p. 30.

[7] Carson and Moulden, p. 69.

STP, Prestone anti-freeze and GLAD bags. First Brands has created a position entitled corporate director for environmental affairs.[8] The position is designed to provide an interface between First Brands and the communities that it interacts with.

To oversee this position a separate division was established, called the Corporate Environmental Affairs Group. The focus of the group is to provide public education and to stay cognizant of governmental legislation and business affairs. The group ensures that First Brands not only complies with governmental environmental legislation, but stays ahead of legislation that could affect First Brands product lines.

The company sponsors the nation's largest recycling campaign, the "GLAD Bag-a-Thon." In 1992, 650,000 volunteers took part in this effort.[9] Such events help showcase corporate dedication to environmentally friendly policies.

Communication of the organization's environmental philosophy, both internally and externally, begins with including environmental views in the mission statement. Mead Corporation requires a statement of environmental position addressing pollution prevention in every business plan. Texas Instruments delineates its environmental policy in the following statement:

> It is TI's policy to provide a safe, healthy workplace and protect the environment while complying with all applicable laws and regulations. TI will remain in compliance with both the letter and intent of all federal, state, and local environmental protection laws; and will go beyond the legal requirements if necessary to protect TI employees, assets, and neighbors from any harmful effects of any manufacturing effluents. Whenever TI policy differs from federal, state, or local laws, the more stringent shall apply.[10]

Another reflection of a company's commitment is the size of the annual budget allocated for environmental protection/prevention (EPP) investment. Companies such as Bayer spend 20 percent of their total manufacturing budget on environmental protection. To put this into perspective, an equivalent percentage of the manufacturing cost is represented by direct labor. Another example of the rising importance of pollution prevention comes from Chevron Petroleum, which is expanding its environmental budget by 10 percent annually throughout the decade.[11]

---

8 Anderson, Eric R. "Going Green: The Corporate Push for Environmental Consciousness." *Business Credit*, January 1992, p. 14.

9 Anderson, pp. 15.

10 Texas Instruments-Internal Document, 1994.

11 Wilsher, pp. 30.

Just as the cost of poor quality was a primary driving force behind corporate quality improvement programs, the cost of inadequate EPP procedures is the key driver behind many advances in corporate waste minimization efforts. The cost of "poor EPP" procedures includes lost sales, disposal costs, fines imposed by regulatory agencies, and lost opportunity for financial recovery from recycling, resale, and reuse.

No one has ever suggested that protecting the environment would be cheap or easy. What *has* been suggested is that it's the responsible thing to do in order to preserve the species.[12]

But pollution prevention is not being touted as just a philanthropic gesture in today's marketplace. It is being treated as a cost minimization strategy and a source of competitive advantage. Measuring the cost savings attributable to effective EPP procedures is difficult at best. Current fines associated with environmental violations can amount to $25,000 per violation per day.[13] While landfill costs can be as low as $50 per drum for the disposal of hazardous wastes, incineration costs can run from $200 to $400 per drum.[14]

The cost of human suffering, lost sales, and environmental restoration is difficult to estimate, but some noteworthy and memorable examples are the Exxon *Valdez* destruction of the Alaskan coastline due to an oil spill, and Union Carbide's damage inflicted on human lives in Bhopal, India, due to chemical leakage. Occidental Petroleum Corporation recently agreed to compensate the state of New York $98 million for the contamination of the Love Canal, with an additional $25 million estimated to cover site maintenance costs in the coming decades.[15]

WHAT IS HAZARDOUS/REGULATED MATERIAL?

A material is considered a hazardous material by the federal government if it demonstrates any of the following characteristics: ignitability, corrosivity, reactivity, or toxicity. A compilation of materials and compounds that fit any or all of these categories is presented in the Resource Conservation and Recovery Act (RCRA). Some frequently cited categories of materials

---

[12] Bergstrom, p. 47.

[13] Rademaker, Ken. "Going, Going, Green." *Occupational Hazards,* September 1991, p. 121.

[14] Kirschner, Elisabeth. "An Anxious Industry Sees New Limits to Its Options: EPA Throws a Curve on Incineration." *Chemicalweek,* August 18, 1993, pp. 23, 24.

[15] *USA Today,* June 22, 1994, p. 3A.

classified as hazardous include chemicals, radioactive materials, flammables, explosives, medical wastes (such as by-products from nuclear medicine and toxic compounds from diagnostic testing), packaging materials, chlorofluorocarbons (CFCs), and hydrochlorofluorocarbons (HCFCs). Many purchasers manage such substances regularly, while some buy them rarely. Regardless of the level of purchasing activity, it is important that all buyers are aware of the special issues related to hazardous materials (HazMats).

Magnitude of the Waste Management Problem

Waste disposal is a growing management concern due to its inherent impact on the environment and the sheer volume of the waste. The following facts are used to help put the magnitude of the problem in its proper perspective:

• Industry generates one ton of hazardous waste for every American citizen annually.[16]

• Every day, American businesses flush more than one billion gallons of fresh water down the toilet.[17]

• By the year 2000, the federal government predicts that solid waste generation will grow by 25 percent.[18]

As is apparent, given the cost of disposing of hazardous waste, the volume involved, and the predicted growth of the problem, developing an effective waste management program is essential to the success of any firm. Purchasing can take the lead role in establishing "green initiatives" if they are not already in place. Every organization generates waste from packaging, office-generated paper, glass, cans, etc., and the purchasing function can serve as the catalyst to eliminate, reduce, or recycle waste products.

Options for Disposal of Hazardous Waste

An analysis of the chemical industry provided by the Chemical Manufacturers Association (CMA) serves as a benchmark for the composition of hazardous waste and the methods selected for waste disposal. A total of 1,910 survey respondents reported the generation of 6.28 billion tons of waste material in 1991, including 315 million tons of hazardous

---

[16] Anderson, p. 14.

[17] Anderson, p. 14.

[18] Sloane, David P. "The Business of Waste Management." *The Environment,* September 1991, p. 18.

wastewater, and four million tons of solid hazardous waste. The preferred methods of treating hazardous wastewater included wastewater treatment for discharge (81 percent), pretreatment for transfer to publicly owned and operated facilities (10 percent), and deep-well injections (9 percent). Solid waste is treated through energy recovery (28 percent), incineration (24 percent), deep-well injection (18 percent), material recovery (13 percent), and landfill use (12 percent), with the remaining waste treated by "other" methods. These findings are illustrated in Figures 7-2 and 7-3.[19]

Four Disposal Methods for Hazardous Waste
There are four basic methods available to purchasing in arranging for the disposal of hazardous material, including: utilizing firms that specialize in hazardous waste disposal; internal processing and conversion of hazardous waste; development of on-site treatment facilities; and contracting an

FIGURE 7-2
Wastewater Treatment Methods

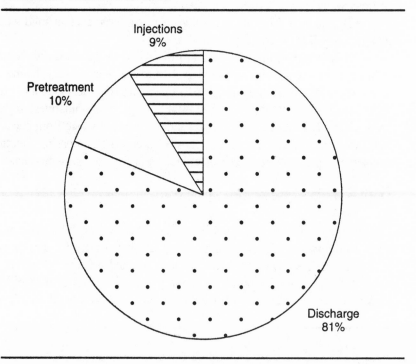

independent commercial waste disposal facility. Subsequent paragraphs discuss each of these options and provide examples.

Utilizing waste disposal companies like Waste Management Corporation, an international organization, for disposing of hazardous/regulated materials is one of the available options. These organizations specialize in providing disposal services, and must adhere to rigid standards to gain approval from the Environmental Protection Agency (EPA). This option reduces the complexity of the disposal decision by taking advantage of the expertise of the full-service supplier. Since the organization that generates the waste is ultimately responsible, care must be exercised in choosing a company that can meet disposal needs properly. These suppliers can be identified by contacting the Environmental Protection Agency or state governments.

The recovery, conversion, or redistribution of hazardous material is another solution offering the potential advantages of waste minimization and lower cost. This solution requires determining whether another part of the firm or another organization can use the converted hazardous material in a

FIGURE 7-3
Solid WasteTreatment Methods

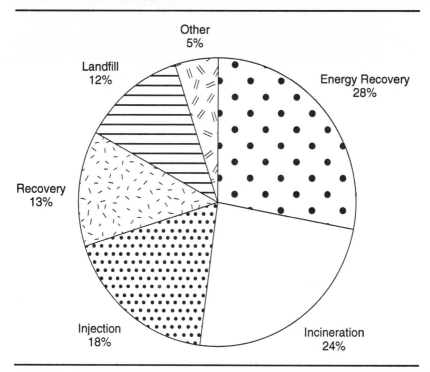

productive fashion. One method to help facilitate this process involves utilizing the services of a "waste exchanger," a firm that specializes in identifying and locating other companies that may have use for various types of hazardous wastes. Purchasing professionals can serve their organization by identifying and developing relationships with suppliers that offer these services.

Veryfine Products, Inc., maker of a wide variety of fruit juices, has adopted a focused corporate strategy to virtually eliminate solid waste produced during its juice manufacturing process. Currently, 90 to 95 percent of all solid waste generated in its manufacturing is recycled. Some of these by-products are transformed into feedstock for animals.[20]

Developing internal, on-site disposal facilities is another option available to firms that possess the technical, physical, and financial capabilities needed for this level of commitment. On-site disposal involves the operation of an internal facility for treatment, storage, reuse, and disposal of wastes. The primary benefit of this option is the degree of control the organization retains over the waste disposal process. While the initial investment is high, the per-unit cost over the long run may actually be the lowest of all the options. One potential impediment is the reaction of the local community, which may exhibit the "not in my backyard" syndrome and protest the development of an on-site waste disposal facility.

An example of on-site capabilities is provided by Herman Miller, Inc., a furniture manufacturer headquartered in Zealand, Michigan, which completed the construction of a waste to energy plant in 1982. This "cogenerator" burns waste products to generate steam, which is then used to power turbines that generate electricity for the plant. This process has cut Herman Miller, Inc.'s solid waste disposal site requirements by 90 percent.[21] It should be noted, however, that this process still puts carbon dioxide back into the air.

The fourth option involves selecting an independent or commercial site for the disposal of hazardous waste. This option differs from the first in that these firms are not considered full-service providers, so the buying firm must be more involved in the process. This involvement may include sorting, packaging, categorizing, and transporting the waste material, and ultimately certifying that the disposal site is in compliance with all government regulations. For purchasing, the availability of sites is an issue, so locating an appropriate facility may require the transportation of waste over long distances. While the transportation issue is the primary deterrent, the primary advantage is the lower cost of disposal.

---

[20] Adams, Samuel. "A Veryfine Approach to Environmental Awareness." *Beverage World,* October 1993, p. 76.

[21] Anderson, p. 14.

Legal Issues Regarding Hazardous Waste

The Federal Occupational Safety and Health Act (OSHA) was passed in 1970. The intent of this law is to protect employees from work conditions that are considered dangerous, hazardous, or unsanitary. Under OSHA, state governments are allowed to institute their own regulations, and they are required to perform inspections to ensure compliance. OSHA policy and procedural requirements are very comprehensive regarding employee exposure to and handling of hazardous material. Failure of firms to provide appropriate safety measures can result in significant penalties.

The Hazardous Communication Standard issued by OSHA requires that workers must be informed about all hazardous materials present in the working environment. This information is communicated through Material Safety Data Sheets (MSDS), which tell employees of physical dangers, safety procedures, and emergency response techniques. The scope of information contained in the MSDS includes the identity of the product, emergency phone numbers, hazardous ingredients, physical and chemical properties, ignitability and explosive hazards, reactivity with other substances, health hazards, and handling and control measures. Purchasing assumes the responsibility for ensuring that MSDS are provided by all suppliers of hazardous materials.

Another component of the MSDS program is the training of employees concerning all aspects of hazardous material procedures. All the training programs must be thoroughly documented, and these records must be kept on file for verification.

The regulation governing the transportation of hazardous materials was passed by the U.S. Congress in 1974, under the Hazardous Materials Transportation act. The U.S. Department of Transportation (DOT) is the federal agency responsible for the enforcement of this act. Hazardous material is categorized into a classification scheme based on a description of the material and its physical properties. From this classification, there are regulations regarding packaging, labeling, transportation, and other pertinent requirements necessary to ensure safe transportation of these materials. Failure to follow the specified requirements detailed by this act can result in the assessment of civil penalties.

The Environmental Protection Agency (EPA) is the federal agency responsible for compliance and enforcement of regulations concerning air, water, hazardous waste, pesticides, and toxic substances. The National Enforcement Investigative Center is part of the EPA. The scope of responsibility is so large that most states have their own version of the EPA. The

EPA is a tremendous source of information regarding available resources and the interpretation of environmental policies.

The DOT and EPA have developed a Uniform Hazardous Waste Manifest, which must accompany all shipments of hazardous waste as specified by the EPA. This document contains the EPA's identification numbers for the organization that generated the waste, the transportation company, and the treatment or disposal facility. It must be signed by the generators, the transporters, and the receiving facility to ensure the traceability of hazardous waste. Current information regarding the policies, orders, and regulations of the various federal environmental agencies is published five days a week in *The Federal Register.* It acts as a comprehensive source of both current and past information concerning hazardous and regulated material requirements.

The Resource Conservation and Recovery Act (RCRA) was passed by Congress in 1976 to protect the public and the environment from potential hazards of waste disposal, to conserve energy and natural resources, to reduce the amount of waste generated, and to ensure that wastes are managed in an environmentally sound manner.

There is a direct linkage between the RCRA and the EPA. The RCRA Implementation Plan (RIP) is the EPA's annual strategic plan for the implementation of its RCRA objectives for the coming year. This plan can be consulted in order to determine exactly where emphasis should be placed within the company in order to comply with government regulations during that year.

It is essential that purchasing professionals are well-acquainted with the applicable federal, state, and local regulations regarding hazardous waste and regulated materials. The cost of non-compliance to the offending firm can be staggering, and it is cost that can be avoided with a sound waste management program. Purchasing can act as the disseminator of critical information to the internal organization.

Ownership and Liability Issues

Hazardous wastes are usually the responsibility of the plant while they remain on-site. But where does the responsibility end? At the manufacturer? With the distributor? Ultimately with the user? It is important to realize that the producer's responsibility regarding the hazardous waste it generates is not alleviated after the waste is disposed of. As long as the material remains hazardous, the producer remains responsible and liable for the consequences. As such, the producer is required to maintain adequate files that allow for full traceability of the material, from "cradle to grave."

This realization has prompted the inclusion of "disposal" in the strategic planning of the product life cycle. "The time has come where it's not enough to develop good, efficient, effective products. The time has come when great thought must go into product life cycle. As in, what happens when what you've made and sold has run its course? What becomes of it? And who should care?"[22] The total cost of introducing a new product must now include its environmental impact and the associated cost to the organization.

The transportation of hazardous waste, as discussed briefly in the section on the Hazardous Materials Transportation Act, involves many liability considerations. Depending on the type and amount of hazardous waste, the producer may be limited to utilizing only carriers approved by the EPA. The purchasing department can develop the internal expertise and do much of the required work, as discussed in the following section, or it can utilize a full-service supplier to handle the scope of issues.

Each shipment of hazardous waste must be accompanied by a carefully completed manifest. The manifest helps in the classification of the material and includes information regarding the producer of the waste, the transporting firm, a description of the materials, and special handling instructions. Title 49 of the Code of Federal Regulations defines the requirements for shipping hazardous materials. The focal point of the regulations is the performance of the packaging. The Office of Hazardous Material Technology within the Department of Transportation groups hazardous materials into one of three groups. The three groups are: Very Hazardous—Packaging Group 1; Medium Hazardous—Packaging Group 2; and Low Hazardous—Packaging Group 3.[23]

It is interesting to note that any citizen may file suit against an organization whose hazardous waste program may be an "imminent hazard or substantial endangerment." In an effort to ensure proper documentation of waste generated, the Pollution Prevention Act of 1990 (PPA) requires that companies must provide data on any recycling, energy recovery, source reduction, or treatment activities involving toxic compounds. This data is then used to construct the Toxic Release Inventory (TRI). The EPA utilizes the Toxic Release Inventory (TRI) Form R to report data on the types and quantities of toxic chemicals released to all environmental media.[24]

---

[22] Bergstrom, p. 46.

[23] Sheridan, John H. "Pollution Prevention Picks Up Steam." *Industry Week,* February 17, 1992, p. 48.

[24] Sheridan, p. 48.

Proactive Strategies for Waste Management

Purchasing's role in the proper disposal of hazardous waste is extremely important in many organizations, because purchasing is either directly responsible or serves in an advisory or resource capacity. The following section presents a variety of proactive strategies dealing with the waste management issue, in which the purchasing function can play a lead role.

Integrated Manufacturing

Minimization of waste prior to manufacturing is the easiest and most efficient way to reduce waste. Integrated manufacturing techniques consider waste management before waste is generated.

The Resource Conservation and Recovery Act, its 1984 Compensation and Liability Amendment, the Pollution Prevention Act of 1990, and the Clean Air Act have all added weight to the need for waste minimization as an integral part of product design. When dealing with integrated manufacturing techniques, it is management's responsibility to ensure compliance and total acceptance of a given policy. The general categories available for management to choose from include source reduction, recovery/re-use, waste exchange, and treatment/destruction/disposal.[25]

Of these methods, source reduction is considered the most economical and environmentally friendly. The extra effort and expense initially required to improve manufacturing processes and product design to reduce the generation of waste is ultimately offset by the effort and cost of hazardous waste disposal. Purchasing can play an active role in this process by becoming involved in new product development teams and process improvement programs. The last method, treatment/destruction/disposal, is the current industry standard, but most companies are undertaking programs that will allow them to develop better alternatives.

Environmental Certification

Green marketing is a tool that can easily be abused in order to increase sales. Companies can easily prey on public sentiment and claim that their products have been manufactured through "new and improved earth-friendly practices." If there is no difference in the actual physical appearance of the product, the consumer is left to assume that the "green improvements" are somehow built into the product.

---

[25] Vajda, Gary, P.E. "Integrated Waste Minimization." *Manufacturing Systems*, January 1992, p. 36.

To help balance the scale, several international standards of "green certification" are currently being developed. These standards follow the quality certification principles of ISO 9000 to an extent. They are not designed to dictate environmental policy. Instead, they are designed to help ensure that companies follow a set of procedures that they themselves designed and agreed to follow. The main objectives of these programs are to help develop credibility and aid the consumer. The European Community (EC) is ahead of the United States in the certification process.

EC Eco-Audit and Management Scheme

The primary objective of this European-based standard was to create a set of guidelines and regulations that industry could use to help monitor itself. The guidelines proposed would establish criteria that industries could use to help ensure a consistent policy of environmental controls. Verification of the standards would be carried out by certified individuals or companies that specialize in this practice.

The guidelines would also require that corporate environmental goals and objectives be published for public disclosure. Those practices that are found to be in non-compliance would have to be modified, or a specific set of goals and procedures would be issued to ensure near-term compliance.[26]

BS7750

Very few goals are attainable unless they are measurable. To this end, the European Community (EC) has established the regulation known as BS7750. BS7750 provides the groundwork for companies to measure their own environmental goals as they are being met. This procedure will also help companies measure the environmental performance of their suppliers.

The standard is also likely to have a major influence on corporate purchasing policies, as big companies require their suppliers to demonstrate that they have effective environmental management systems and audits.[27]

The Green Cross Certification

The Green Cross is an independent assessment of environmentally friendly claims made by a company. This independence is what gives the Green Cross its credibility. Four major grocery chains and 200 products currently carry the Green Cross Certification. As with ISO 9000 and other standards of compliance, the number of participants is increasing rapidly. Along with

---

[26] Rowson, David. "Going Green." *Management Accounting,* January 1994, p. 20.

[27] Rowson, p. 20.

the Green Cross Certification System comes a great deal of reading material designed to educate consumers. The material includes informational brochures and explanations of the program's intended goals.[28]

The Green Report Card

Stanley Rhodes, president of the Oakland-based Scientific Certification Systems (SCS), has introduced a method of monitoring environmentally destructive processes. The Green Report Card is designed to tell the company exactly what it is doing, and not to dictate what the company *should* be doing. This process can be used as a method of monitoring progress, and verifying waste management procedures, and as input into supplier selection criteria.

The report card uses a bar scale to convey information easily and effectively. The areas assessed include resource depletion, energy use, air pollution, water pollution, and solid waste generation and disposal. Each category lists specific types of pollutants and includes a corresponding bar scale that indicates the exact quantity of each type of pollutant that is being generated. The report card then details this information on an easily readable scale, which rates the company from good to poor on a continuum.

An "800" number is provided so the consumer can learn more about how the categories were established and how each category is verified. The strength of the report card strategy is its ability to provide detailed information on the product's environmental impact in a minimum amount of time.

Emerging Standards for Environmental Management Systems (EMS)

EMS is another third-party verification system to ensure compliance with a set of procedures and guidelines that a company adopts in order to track its environmental performance. As with ISO 9000 and the other certification methods previously mentioned, this certification requires that a qualified third party conduct an independent inspection to ensure compliance with previously outlined procedures.

The eventual goal is to make EMS an international standard.

The long-range objective is to have a dual set of standards that can be monitored concurrently to help minimize the overhead expenditure associated with the certification processes.[29]

---

[28] Sansolo, Michael. "Going Green: Three Ways to Build Trust." *Progressive Grocer,* February 1991, p. 46.

[29] Marcus, Jeff. "Trends Purchasing and the Environment." *NAPM Insights,* April 1994, p. 40.

Responsible Care

The chemical industry responded to public concerns that it was damaging the environment by adopting an industry-wide reporting mechanism. In order to become a member of the Chemical Manufacturers Association (CMA), a chemical company must first follow the guidelines. The program was initiated in 1988 and is designed to increase corporate accountability to the public sector.[30]

Green Purchasing

Beyond recycling lies source reduction. By preventing waste before it occurs, companies can incur a greater cost saving than previously possible. Preventing waste at its source helps to eliminate waste at its earliest stage. The key to this concept lies in green purchasing.

Organizations must have procurement programs in place ensuring that environmental objectives are met. Suppliers must be rated on their ability to conform to green standards. Criteria that are measurable and enforceable must be developed. Continuous feedback is essential to such a program. Suppliers that are non-compliant with previously agreed upon criteria should undergo a series of warnings and penalties until the supplier meets conformance standards, or until that supplier can be replaced by a new, qualified supplier.

One of the first and easiest green purchasing methods a company can employ involves reducing the amount of packaging required for shipment of its inbound materials and finished goods. Packaging materials range from annoying to environmentally damaging. Some of the older foam packaging materials that were so widely used only a few years ago are non-biodegradable, and will remain in landfills for thousands of years.

The easiest solution to this problem is to use packaging made from recycled materials. However, according to Robert Rothfuss, vice president of marketing for Buckhorn, Inc. (manufacturers of reusable containers), "A more sophisticated approach, and one with more impact potential, is shipping in reusable packaging."[31] The problems associated with this approach are logistical in nature, and could possibly prevent the idea from materializing. Fuel costs would be incurred as shippers are forced to return the containers to their point of origin. Such a progressive solution would require a

---

[30] Marcus, p.40.

[31] Andel, Tom. "Don't Recycle When You Can Recirculate." *T & D,* September 1991, p. 68.

high degree of coordination between the consumer and the supplier. Returnable containers are currently utilized in the automotive industry to handle bin stock, component parts, and sub-assemblies.

In order to address some of the problems that could prevent a successful reusable container program, Mr. Rothfuss suggests that the following measures should be undertaken: ensure cooperation among all the participants in the program; establish an economical method of returning empty containers to their source (i.e., pallet pools); utilize a small number of standardized containers; develop container tracking systems; and establish frequent shipments to expedite container investment payback.[32]

A great deal of this effort requires that engineers experiment with new materials on a case by case basis, in order to pinpoint areas where elimination or change can occur. Currently, wood is being replaced by corrugated paper, laminated paper bags take the place of poly bags, paper pulp molds replace Styrofoam molds, and new adhesives replace stretch wrap films.[33]

Corrugated cardboard boxes were used to ship almost everything in industrial supply chains, and are one of the most common sources of waste. Methods to increase overall stacking strength allow the amount of material used in box construction to be reduced. Such efforts require a great deal of discipline and employee participation.[34] The bottom line cost accounting returns for such efforts is not yet fully known. However, a proactive stance to environmental awareness is clearly the best policy.

Success Stories from Industry

- The Steelcase Co., a Grand Rapids, Michigan, multinational corporation that manufactures office furniture, has gone to great lengths in its efforts to minimize waste. The company's goal is to eliminate all trips to the landfill. In order to support this goal, the company has started a program that taps every employee at every level for suggestions that could help improve its manufacturing process. Suggestions have included new machining improvements, packaging source reductions, and recycling/reuse programs.[35]

- In 1987, AT&T's printed wiring board factory, located in Richmond, Virginia, began an intensive, 10-year effort to reduce

---

[32] Andel, p. 72.

[33] Andel, p. 68.

[34] Andel, p. 68.

[35] Andel, p. 68.

toxic wastes. Results were almost immediate. By 1990, the amount of chlorinated solvents released to the atmosphere had been reduced to 40,000 pounds per year, down from 3,000,000 pounds per year.[36]

- AT&T's corporate objective was to stay ahead of the EPA regulation requiring a 50 percent reduction in the emissions of 17 toxic chemicals by 1995. AT&T Chairman Robert E. Allen notified the EPA that, by 1995, the company should achieve a 95 percent reduction in toxic waste emissions.[37] Only with similar support from the very top levels of management can programs such as these succeed.

- Northern Telecom has developed a process that eliminates the use of CFCs to remove excess solder from circuit board assemblies. The new technology will save an estimated $50 million and will prevent the release of 9,000 tons of CFC by the year 2000.[38]

SUMMARY

Most large corporations have developed written environmental objectives and policies. These policies provide a measure for the companies to use internally, as well as for investors/consumers to use in judging the company. Corporation's sense of environmental awareness is heightened, and they are being further educated through the competitive processes of the open market, such as the environmental certification process. This process adds credibility to environmental claims, while at the same time making products that lack the relevant seal of approval appear less competitive.

The current environmental movement is not a trend. Instead, it is a market force driving and influencing entire segments of the economy. A full-blown, capital-intensive effort to change business processes throughout various industries is occurring. Those companies that have chosen to lag behind industry leaders in this field find themselves in the unenviable position of trying to catch up. The pressure being applied from consumers, governments, and competing firms is far too great to be ignored.

---

[36] Sheridan, p. 37.

[37] Sheridan, p. 37.

[38] Sheridan, p. 48.

KEY POINTS

1. Purchasing's role in corporate social responsibility involves the proactive investigation and implementation of environmentally friendly strategies, policies, and procedures.

2. Green practices are driven by market demands rather than government legislation.

3. Implementing a green strategy takes the leadership and support of upper management and the dedication of corporate resources.

4. The cost of poor environmental protection/prevention procedures for an organization can be staggering, resulting in lost sales, disposal costs, fines, and lost opportunities for financial recovery, resale, and reuse.

5. A material is considered hazardous if it demonstrates ignitability, corrosivity, reactivity, or toxicity.

6. There are four basic methods for the disposal of hazardous wastes, including: utilizing firms specializing in the disposal of hazardous waste; internal processing and conversion of hazardous waste; on-site treatment facilities; and contracting an independent commercial waste disposal facility.

7. Government regulation directed at the handling and disposal of hazardous waste includes the Occupational Health and Safety Act (OSHA), the Resource Conservation and Recovery Act (RCRA), and the Hazard Materials Transportation Act. The Environmental Protection Agency (EPA) is responsible for the enforcement of this legislation.

8. The producer of hazardous waste is ultimately responsible for the impact this material has on the environment.

9. Proactive strategies available to the purchasing manager include integrated manufacturing, environmental certification, EC eco-audit and management scheme, BS7750, the Green Cross certification, the Green Report Card, emerging standards for environmental management systems, responsible care, and green purchasing.

10. Green purchasing includes preventing the generation of waste material. It is referred to as source reduction, which involves strategies aimed at preventing the creation of waste.

SUGGESTED READINGS

Adams, Samuel. "A Veryfine Approach to Environmental Awareness." *Beverage World*, October 1993, pp. 76-78.

Andel, Tom. "Don't Recycle When You Can Recirculate." *T & D,* September 1991, pp. 68-73.

Anderson, Eric R. "Going Green: The Corporate Push for Environmental Consciousness." *Business Credit,* January 1992, pp. 14-17.

Baker, R. Jerry, C.P.M. "Our Changing World: The Social Responsibility of Purchasing." *NAPM Insights,* February 1993, p. 2.

Bergstrom, Robin P. "On Giving Due Thought to Afterthought" *Production,* November 1991, pp. 46-47.

Bierlien, L.W. *Red Book on Transportation of Hazardous Materials.* New York: Van Nostrand Reinhold, 1988.

Carson, Patrick, and Julia Moulden. "Green is Gold." *Small Business Reports,* December 1991, pp. 68-71.

Charles, Robb. "Zero Loads to Landfill." *Biocycle,* July 1991, pp. 62-64.

Kirschner, Elisabeth. "An Anxious Industry Sees New Limits to Its Options: EPA Throws a Curve on Incineration." *Chemicalweek,* August 18, 1993, pp. 23-24.

Leaversuch, Robert D. "Plasticscope: In Many Ways, the Solid-Waste Issue Now Cuts into Sales of Virgin Resin." *Modern Plastics,* February 1991, pp. 16-18.

Leavitt, Paul. "Love Canal." *USA Today,* Wednesday, June 22, 1994, p. 3A.

Magrath, Allan J. "The Marketin' of the Green." *Sales & Marketing Management,* October 1992, pp. 31-32.

Marcus, Jeff. "Trends Purchasing and the Environment." *NAPM Insights,* April 1994, p. 40.

Rademaker, Ken. "Going, Going, Green." *Occupational Hazards,* September 1991, pp. 121-23.

Rowson, David. "Going Green." *Management Accounting,* January 1994, p. 20.

Sansolo, Michael. "Going Green: Three Ways to Build Trust." *Progressive Grocer,* February 1991, pp. 45-46.

Schlossberg, Howard. "Hardware Industry Understands Importance of Going Green." *Marketing News,* October 11, 1993, p. 10.

Sheridan, John H. "Pollution Prevention Picks Up Steam." *Industry Week,* February 17, 1992, pp. 36-48.

Sloane, David P. "The Business of Waste Management." *The Environment,* September 1991, pp. 18-19.

Texas Instruments, Internal Document—Waste Disposal Policy and Guidelines, 1994.

Vajda, Gary, P.E. "Integrated Waste Minimization." *Manufacturing Systems,* January 1992, pp. 36-40.

Wasik, John. "Market Is Confusing, but Patience Will Pay Off." *Marketing News,* October 12, 1992, pp. 16-17.

Wilsher, Peter. "The Feeling Grows That Going Green Is Good for Business." *Management Today,* October 1991, pp. 27, 30.

CHAPTER 8

PURCHASING PLANNING AND ACQUISITION STRATEGY

What is planning? In general, planning is the act of determining a proposed or intended course of action or direction to be taken in achieving future goals and objectives. In business, planning usually entails developing a detailed description of the goals of the organization, plans or blueprints, and a timetable for attaining those goals. The planning process starts by determining the mission or purpose of the organization, followed by developing company-wide goals and the means of attaining those goals. This includes setting tactical and operational goals and plans for all levels of the organization. Goals can be called the ends that the organization desires to reach, and its plans are the means that it will use to achieve those ends. The planning process should be continuous. Current decisions should take the future into consideration, and actual results should always be measured against the established expectations.

THE PLANNING HORIZON

Three levels of planning exist—strategic, tactical, and operational. At the strategic level, company-wide goals and plans are defined in broad terms. Strategic goals indicate where the organization wants to be in the future, and strategic plans outline how the organization intends to get there. Peter Drucker, a leading management consultant and author, suggests developing strategic goals for eight content areas: market standing, innovation, productivity, physical and financial resources, profitability, managerial performance and development, worker performance and attitude, and public responsibility.[1]

---

[1]Drucker, Peter F. *The Practice of Management.* New York: Harper & Brothers, 1954, pp. 65-83.

The tactical plan takes the strategic plan to a more detailed level and outlines the expectations of each major division and department. The operational plan details the expectations for individuals, departments, and work groups.

Long-term, intermediate-term, and short-term time horizons are usually associated with the planning process. Long-term planning extends five years or beyond, and deals with strategic goals and plans in the broadest sense. Intermediate-term planning usually covers one to five years, may be general or specific, and addresses tactical objectives. Short-term planning covers one year or less, and deals with operational objectives for specific departments and individuals. Operational objectives are precise and measurable short-term goals.

Linking the Three Planning Levels

The three planning levels are linked in a hierarchical fashion. The success or achievement of lower-level plans enables the achievement of higher-level goals. In other words, the operational objectives are the means of reaching the tactical or intermediate goals, and the intermediate objectives are the means of attaining the long-term strategic goals. This is shown in Figure 8-1.

Purchasing and the Planning Process

Purchasing is in a unique position. It has access to internal information because of its relationship to the other functional areas in the organization. It also has access to external market information through its suppliers, and through its knowledge of the marketplace and economic trends. Purchasing or supply management should play a role in the strategic planning process of the organization. The strategic dimension of supply is also addressed in the first book in this series.

D. Larry Moore, Ph.D., C.P.M., and president of Honeywell Space and Aviation Systems, says purchasing is "your first perspective on the outside world. Jobs prior to purchasing were internally focused. The biggest personal benefit was interfacing with the external environment and learning the dynamics of the different industries because of supplier contacts. It continues to amaze me how many people underestimate the tremendous and obvious leverage purchasing provides."[2]

---

[2] Murphree, Julie. "Top Brass Polished in Purchasing." *NAPM Insights,* February 1991, p. 25.

FIGURE 8-1
Strategic Contribution of Supply

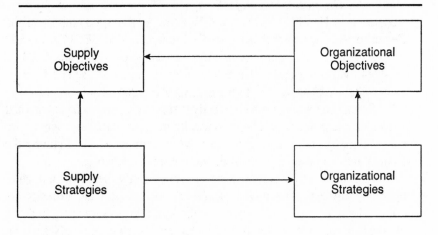

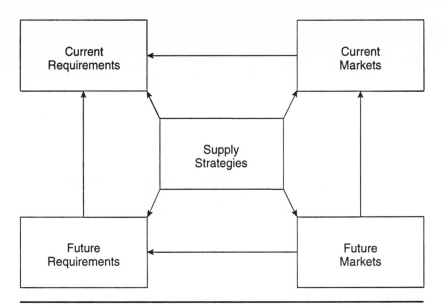

Source: Michiel Leenders and Anna Flynn, *Value-Driven Purchasing: Managing the Key Steps in the Acquisition Process,* Burr Ridge, IL: Richard D. Irwin, Inc., 1995, p. 9.

Issues in Purchasing Planning

Support of Organizational Objectives

According to Harold E. Fearon and William A. Bales, co-authors of the Center for Advanced Purchasing Studies' report entitled *CEOs/Presidents' Perceptions and Expectations of the Purchasing Function,* the CEO or president will always ask, "What did you do to improve this period's profits or to have a positive effect on the company and shareholders' value?"[3]

Purchasing strategies and objectives must be linked to organizational strategies and objectives. The individual purchaser must keep one eye on the strategic goals of the organization while simultaneously attaining the tactical and operational goals of the purchasing function.

According to D. Larry Moore, Honeywell's president and chief operating officer, "The really successful procurement departments are doing a lot of things today they probably didn't do ten years ago, in terms of working on such things as concurrent engineering and early supplier development."[4]

In keeping with the hierarchical nature of the planning process and the linkage between the levels of planning, purchasing decisions must consider the wider context of the entire organization and how that organization interfaces with the environment. This is shown in Figure 8-2.

For example, a university president is faced with many constituencies—faculty, students, regents, legislators, and the taxpayer. Earl Whitman, director of purchasing and risk management at the University of Oklahoma, explains the impact this broad constituency has on a purchase decision:

> Any major purchase award decision or program must consider the impact on each of these constituencies. The purchasing decision cannot be made solely in a narrow purchasing perspective, but in the wider perspective of the entire university. A plan for selling, justifying, and communicating major purchasing actions to the impacted groups is vital to acceptance, because it will help to minimize potential problems for the president and smooth the way for approval. The perspective of purchasing can then be one of "problem solver" rather than "problem creator."[5]

---

[3] "Winning with Upper Management," *NAPM Insights,* October 1993, p. 33.

[4] Murphree, p. 25.

[5] "Winning with Upper Management," p. 33.

FIGURE 8-2
Three Perspectives on Supply Contribution to Organizational Objectives and Strategies

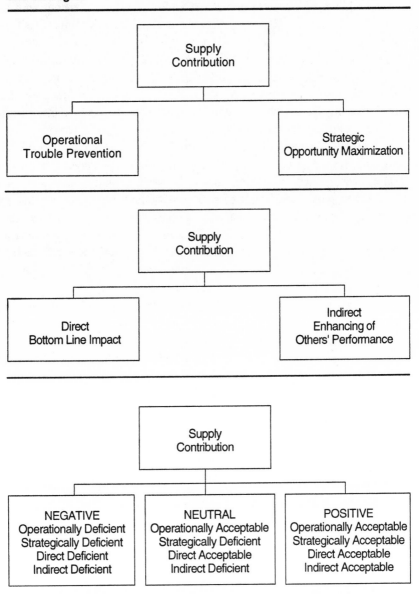

Source: Michiel Leenders and Anna Flynn, *Value-Driven Purchasing: Managing the Key Steps in the Acquisition Process,* Burr Ridge, IL: Richard D. Irwin, Inc., 1995.

Profit or Revenue Planning

Although many organizations feel that total emphasis on profits or revenues ignores the broader purpose of the organization, most organizations set some sort of profit/revenue objectives. Specific profit targets might involve achieving a pretax margin on sales of 10 percent or a return on invested capital of 15 percent. Profit and revenue objectives usually flow from the top down. An overall profitability level expressed in terms of rate of return or profit margin is established by corporate managers. These goals are then translated into objectives for specific business units, which are further specified into the selling prices and/or volumes necessary to generate the desired return.

Profit plans are affected by a number of factors, including product volume, selling price, product mix, product design changes, hourly and salary wage rates, purchased materials, price changes, new plant and equipment purchases, levels of productivity, and improved systems and methods. Management has varying degrees of control and, hence, forecasting ability over these factors. The four factors most easily forecasted because of the high degree of internal control are: hourly rate changes from labor contracts; new plant and equipment purchases; estimated productivity levels; and improved systems and methods.

Since material prices and design changes are heavily influenced by external factors, forecasting is more difficult. Price indices, seller's feedback, and past experience provide useful information for forecasting materials prices. Refer to Chapter 2 for a more detailed discussion of forecasting. Design changes affect the cost of manufacturing, purchasing, and the product volume based on consumer acceptance. Consumer preferences, level of economic activity, competitors' decisions, and government actions may all influence selling price, volume, and product mix, thereby making these items difficult to predict and control.

Commodity Plans

Purchasers can use economic forecasting information such as that found in the Report on Business (ROB), available monthly in *NAPM Insights,* to predict the price and availability behavior of commodities. For example, if the ROB indicates that a commodity is in short supply, the purchaser may assume that pressure will be on to increase the price. The buyer may then need to lock in prices and assure supply.

Contingency Plans

Contingency plans outline a company's planned response to unforeseen events, such as recession, inflation, material shortages, technological developments, or safety incidents. By developing responses to worst-case scenarios, the company management aims to minimize the impact should any of these events occur.

Decision Making

The ability to make good decisions affects the outcome of the plan. Decision making consists of three steps: defining the problem, developing and evaluating alternatives, and implementing the chosen alternative. Problem definition is critical, because if the problem is misdiagnosed, it can't be properly solved. Symptoms are often identified as problems, and treatment of the symptom does not resolve the problem. For example, high stock out levels may be diagnosed as the problem, while in fact it is only a symptom of frequent production line downtime, which may really be caused by a supplier quality problem. Until the problem is accurately diagnosed and all affected parties agree on the diagnosis, it is pointless to try to develop and evaluate alternatives.

Generating alternative solutions is actually a two-step process. First, it is good to brainstorm as many alternatives as possible in an open, friendly environment in which creative thinking is fostered. The ground rules for brainstorming are as follows: all possible ideas are listed; no evaluation is performed; and no negative feedback is given. After all ideas are on the list, the evaluation process begins. Evaluation of alternatives includes gathering information and performing cost-benefit analysis to narrow the list down to the best alternative under the circumstances. Both qualitative and quantitative analysis should be considered.

Cost-benefit analysis and opportunity cost analysis are required before an alternative may be selected. Cost-benefit analysis involves determining the costs of the alternative compared to the likely benefits of implementation. For example, a customer service level of 95 percent may require an inventory of $150,000; however, increasing service to 98 percent may increase inventory to $450,000. To make this option worth the expense, the additional revenue generated would have to cover the $300,000 increase in inventory. Making this decision requires input from a number of individuals in the organization. The usefulness of each person's input affects the success of the chosen alternative.

Opportunity cost analysis identifies the opportunity lost by choosing one alternative over another one. Opportunity cost is equal to the cost of capital, or the rate of return that would have been realized if the capital had been invested in the other alternative. In the previous example, choosing to stay with a 95 percent service level means losing the sales revenue the firm would have generated at a 98 percent level. The benefit is a decrease of $300,000 in inventory.

The final step in selecting an alternative is selling that alternative to other affected parties in the organization. Some internal selling will occur in the information gathering phase, but it is necessary to determine whose approval and what type of approval is needed before implementing the alternative. Implementation translates an idea into action. Depending on the issue, some level of management commitment is needed to ensure organizational support in the form of budgeted resources and policy and procedural changes.

Periodic Review
All purchasing planning should be a continuous process. The plan itself is a living entity, not something that is engraved in stone and stored away for safekeeping. It should evolve to reflect the opportunities and challenges of both the internal and external environments. Building feedback loops into the formal planning process helps ensure that the purchasing plan is dynamic, rather than static. The process should include periodic reviews of progress toward milestones and goals.

Advanced Acquisition Planning
Tying purchasing's goals and strategies to those of the organization is critical if the purchasing function is to be recognized as a player on the strategic team. The earlier that purchasing is involved in the strategic planning process, the greater the opportunity for direct and indirect impact on the bottom line.

Some of the issues the purchaser should consider when developing acquisition plans are: whether the product is new or existing; the dynamics of the market for the item; the general availability of the product; the possibility of implementing a JIT process for purchasing the item; the costs of carrying the product in inventory; the readiness of possible suppliers to produce the needed product; and projected lead time.

Linking Purchasing Strategies with Forecasts

Forecasts of Volume
Based upon projected needs, there are several contract types, supply agreements, and buying strategies that may be appropriate. These include hand to mouth purchase, systems contracts, and multi-year supply contracts, and they are tools by which the buyer relates forecasts of need to supply market conditions.

Annual Requirements

Marketing studies and forecasts, coupled with historical usage data, allow purchasing to forecast the organization's material needs. This determines the annual requirements of everything from service and operating materials to capital equipment. For planning purposes, annual requirements are often determined by aggregating the demand on the commodity level, then disaggregating this information to the individual item, or component level, for placing requisitions.

Life of Product or Part

At times, it is useful to forecast requirement volumes not by month or year, but over the entire life of the end product or service. This part or product life projection can form the basis of life-of-product or service contracts. With the continuing trend toward shorter life cycles, the utilization of this planning technique is expected to grow. For example:

> The Austad Company, a cataloger of golf equipment, used to forecast solely on units sold during the prior year's season. Since Austad started adjusting forecasts up or down depending on life cycle expectations for each product, forecast accuracy has improved.[6]

Buying Strategies

Timing of purchases is among the strategies available to purchasing and materials managers. Use of forecast data can assist in choosing the appropriate option from those listed in the following section:

Hand To Mouth Buying

This is a short-term strategy, typically described as purchases of requirements for a time period up to two or three weeks from the date of purchase. This strategy might be employed in falling markets, in which buyers wish to take advantage of decreasing prices with each successive purchase. When facing cash flow constraints, or dealing with goods that are either perishable or subject to rapid technological change, this approach may also be appropriate.

Buying To Requirements

Advance purchases for use in the next three weeks to three months might be classified as "buying to current requirements." This practice is probably

---

[6] Wisner, Joel D. "Forecasting Techniques for Today's Purchaser." *NAPM Insights,* September 1991, p. 23.

the most common, as it assures supply while avoiding excessive inventory carrying costs.

Forward Buying Versus Speculation

Frequently, conditions such as potential supply constrictions or inflationary markets cause purchasing managers to buy more of a product than is required. This practice, called forward buying, serves to protect the organization from shortages, or to delay the impact of rising prices. The trade-off is, of course, increased inventory carrying costs. The purchasing manager must evaluate the trade-off between inventory carrying cost increases and the risk of supply constriction or increased prices when using this strategy.

> The Ethyl Corporation, a petrochemical and pharmaceutical intermediary, forecasts raw materials needs and prices in their purchasing department. A quarterly outlook based on major economic indicators is created by an economist at Ethyl to aid in the purchase of materials. According to Steven Moore, director of commercial services, "During the period leading up to the conflict in the Persian Gulf, we saw a major increase in petroleum-based raw material prices; we predicted this price increase and we were able to benefit from advance purchases of these materials."[7]

Speculative buying refers to purchases made not for internal consumption, but rather with the intent to resell at a later date for a profit. These speculative goods may be the same as those purchased for consumption, but quantities purchased will be in excess of current or future needs. The fundamental intent is to take advantage of expected increases in price to profit from the resale of the goods.

Volume Purchase Agreements

When significant quantities of specific products or commodities are needed, these requirements may be met through volume purchase agreements. The primary objectives of these agreements are to assure supply and to maximize purchase leverage. Depending on the duration of the demand, these agreements may be either short-term or long-term, and they may take many forms. Depending on the circumstance, these agreements range from specific descriptions of volumes to be purchased to very nebulous commodity descriptions. These agreements may be called blanket orders.

---

[7] Wisner, p. 23.

Life of Product

It may be desirable to award contracts to suppliers of raw materials or components for the life of the end-product, sometimes referred to as evergreen contracts. If duration of need is limited, it may not be cost-effective to rebid. Familiarity with needs of the buyer, the use of the item, and special supplier capabilities are other reasons for this type of contract. Often, this sort of agreement is developed between a buyer and a supplier that have a long collaborative history, which may include such activities as joint engineering of the components to be supplied.

Just-in-Time

Just-in-time (JIT) manufacturing is more a philosophy of doing business than a specific technique. The JIT philosophy focuses on the identification and elimination of waste wherever it is found in the manufacturing system. The concept of continuous improvement becomes the central managerial focus. Several of the more highly publicized results of JIT implementation are the initiation of a "pull" system of manufacturing (match production to known demand); significant reductions of raw material, work-in-process, and finished goods inventories; significant reductions in through-put time; and large decreases in the amount of space required for the manufacturing process.

The greatest improvement for a company implementing JIT, however, is usually in the area of quality. If there is little or no raw material inventory, then incoming raw material and components must be of impeccable quality or manufacturing will cease. Similarly, each intermediate manufacturing step must yield high-quality output or the process will stop.

JIT philosophy focuses on the elimination of waste wherever it is found in the business system, which includes the supplier. The aim is to reduce waste and cost throughout the entire supply chain. If a manufacturer decides it will no longer carry raw material inventory, and that henceforth its suppliers must carry the inventory, this does not reduce supply chain cost. It only transfers costs from one link in the supply chain to another. While those additional inventory carrying costs may be borne in the short term by the seller, they eventually must be passed on to the buyer in the form of higher prices.

One of the most often cited reasons for difficulty in the implementation of JIT is a lack of cooperation from suppliers, due to the changes required of a supplier's system. In addition to changing from traditional quality control inspection practices to the implementation of statistical process control, the supplier is asked to manufacture in quantities that may differ from the usual lot sizes, and to make frequent deliveries of small lots with precise timing. Additionally, the supplier is normally required to provide the buyer access to its master production planning system, shop floor schedule, material requirements planning system, managerial system, and financial statements.

Under JIT, close and frequent buyer-supplier communication is essential. Suppliers are given long-range insight into the buyer's production schedule. Often, this look ahead spans a dozen weeks or more, and the schedule for the nearest several weeks is frozen. This allows the supplier to acquire raw materials in a stockless production mode, and to supply the buyer without inventory build-ups. Suppliers provide daily updates of progress, production schedules, and problems. Clearly, purchasers and suppliers must cooperate and have a trusting relationship in order to convert supply chains to JIT operations.

Supplier selection, single sourcing, supplier management, and supplier communication become critical issues for purchasing and materials managers in implementing JIT. Critical issues in JIT supplier selection include quality control methods, supplier proximity, manufacturing flexibility, and reliability. JIT firms and their suppliers generally develop close collaborative relationships supported by long-term, single-source contracts. The concept of partnering is often applied to the JIT buyer-supplier relationships.

Following supplier selection, careful supplier performance measurement and management often lead to supplier certification—a designation reserved for those suppliers whose quality, on-time delivery, and reliability are proven acceptable over long periods of time.

> At the Ethyl Corporation, purchasing managers are responsible for knowing supply lead times. Since there is little room for storage of materials for the petrochemical and pharmaceutical intermediary, the company operates in a JIT mode to accommodate weekly manufacturing schedules. Using experience, sales forecasts, and historical usage as a basis, monthly or quarterly forecasted requirements are easier to predict. A rolling four-quarter forecast for material prices is used, and forecasts are tracked for accuracy.[8]

The functioning of the purchasing department has significantly changed under JIT from the processing of orders to a focus on supplier selection and long-term contract negotiation. Many times, these close communications are supported by electronic data interchange (EDI) capabilities to facilitate the timely and accurate transmittal of information.

---

[8] Wisner, p. 23.

Consignment

Inventories that are owned by the supplier but stored at the buyer's facility are said to be "on consignment." These goods are billed to the buyer only after they have been consumed.

At first glance, this practice seems to be advantageous to both buyer and supplier. The supplier has an assured sale, while the buyer has the security of on-site inventory without inventory investment. There are, however, potential problems with this procedure. For example, even though consignment inventory is stored at the buyer's warehouse, it is still owned by the supplier. As a result, the supplier may want to remove some of the items to sell to another customer, while the buyer is counting on those items to cover his or her own requirements.

The fact that the buyer does not invest in consignment inventory, but only pays as it is used, does not relieve the buying firm of paying for the remainder of the inventory carrying costs. The supplier builds these costs into the price of the item. Additionally, it is still necessary to provide a secure facility for consignment inventory, to care for it, and perhaps to track it.

Implementation of Buying Strategies

There are several common techniques for the implementation of purchase strategies based on forecast data. The following section will address the techniques of spot buying; hedging; dollar averaging; contracting; escalation and de-escalation clauses; multi-year contracting; line of products contracts; future delivery; options; and buying capacity.

Spot Buying

Spot buying is the practice of buying a commodity on the "spot," or on the open market at current market prices. This is common in the procurement of raw materials.

Hedging

Hedging typically involves the sale of a future contract to offset the purchase of a cash commodity. It can also involve the purchase of a futures contract to offset the sale of a cash commodity that is included as part of an end-product sold by the firm. A firm simultaneously enters into two contracts of an opposite nature—one in a cash market and one in a futures market.

Hedging is used, for example, to safeguard profit margins when a sales contract with fixed prices and delivery over an extended time period is negotiated, but the purchase of the raw material used to produce the item is postponed. If material prices were to rise between the time of the order and the actual purchase of raw material, the profitability of the entire contract could be jeopardized. Hedging is also used to protect inventory values in a declining market. If declining raw material prices cause finished product prices to fall, the firm will lose money due to the relatively higher price paid for raw materials at an earlier time. With a futures contract, the firm can insure itself against price fluctuations in the raw materials.

As an example, assume that a firm established a contract in April to supply generators that will be delivered to a customer in December. The generators contain 25,000 pounds of copper, which were included in the bid price at $1.04 per pound. Because of production lead time, the copper will not be needed until October. To guard against a possible increase in the price of copper, the purchasing manager buys an October futures contract for 25,000 pounds of copper at $1.10 per pound, which is the April price of copper for delivery in October. If the spot (cash) price rises, typically the futures price will rise by a similar amount. In our example, the spot price rises from $1.04 to $1.10 by October, and the October futures price rises to $1.16 per pound. Thus, the futures contract can be sold for a $0.06 per-pound profit ($1.16 October futures price—$1.10 October spot price). This offsets the rise of $0.06 in the price of spot copper when the purchasing manager buys the copper in October.

Dollar Averaging
When purchasing a commodity or component over time, the value of the items in inventory is an average based on the buying of quantities at different times for different prices. For example, if 10,000 pounds of copper are bought at $1.04 per pound, 7,000 pounds are bought at $1.07, and 13,000 pounds are bought at $1.09 per pound, the averaged price of the commodity would be $1.064 per pound [(10,000 x 1.04) + (7,000 x 1.07) + (13,000 x 1.09)/ 30,000]. The averaging process dampens the departure of short-term price fluctuations from long-term averages.

Contracting
Rather than selecting a supplier and placing an order each time a requirement occurs, most firms today select suppliers for longer-term agreements. These contracts typically involve products whose consumption represents a significant dollar value on a continuing basis. Contracts may also be written to cover families of products or classifications of products, such as office materials or electrical supplies. These agreements may take many different forms, including multi-year contracts, life-of-product contracts, future delivery agreements, or perhaps even options on products or supplier's capacity.

Escalation and De-escalation Clauses

In any contract of long-term duration, both parties are at risk from input cost changes. A buying firm would not want to agree to a fixed price contract, only to watch the worldwide price of the raw material fall significantly. The supplier would likewise be harmed by entering into a long-term agreement, only to find labor or raw materials prices rising. The mechanism for the sharing of such risk is called a price change clause, or an escalation clause, which provides for price adjustments based on indexes that reflect material and labor cost changes. It is an equitable method of assuring that both parties share the risk of economic changes beyond their control. The fact that many price change clauses do not share the risk or cost change burdens equally illustrates that these clauses must be carefully crafted. Risk sharing is a fundamental building block of partnering relationships.

Multi-Year Contracting

In today's business environment, buyer and supplier firms are entering into close collaborative relationships based upon mutual benefits. Buyers are reducing supply bases to include only suppliers judged to be superior. In such cases, it is likely that the firms will enter into agreements that clarify mutual expectations, and that typically span several years. After very rigorous supplier selection procedures, some purchases are now entering into contracts with 6- to 10-year durations. The critical activity is the selection of such long-term partners/suppliers—this often takes from 6 to 12 months. After investing this time and effort, however, there should be no need to repeat the selection process until the long-term contract expires.

When formulating long-term agreements, it is important for purchasing and materials managers to be sure that provision is made for specific, periodic performance and satisfaction reviews by both parties. Problems must be addressed in a timely and orderly fashion. As with any long-term agreement, provisions for dissolution should be agreed upon prior to contract signing.

Life of Product or Service Contracts

One of the long-term contracting options is to agree that suppliers will provide materials, components, or services not for a specific time period, but for the entire life span of the product. Life-of-product contracts can be useful in developing cooperation between a buyer and a supplier for products that have short life cycles, that are highly complex, or for which technology is changing rapidly. For longer-term situations, this contract form encourages the supplier to invest in necessary training, equipment, or technology without needing to recover the investment during a one-year contract life.

Product or service standardization is a strategy that facilitates this form of agreement.

Future Delivery

A purchasing manager's firm may have major requirements only on a sporadic basis, or may be in the process of developing a new product. In such cases, a buyer may want to ensure that goods will be available from a particular supplier within a particular time frame. If this is the case, a future delivery contract can be negotiated to ensure that productive capacity is reserved. Sole source suppliers of items such as capital equipment may fall into this category as well. As discussed previously, price change clauses are critical to these contracts.

Options

A purchase option is the right to purchase something under agreed terms for a specified period of time. When the specified period of time expires, so does the right to purchase the goods at the agreed upon price. Such rights are often granted for a negotiable fee, which is forfeited if the right to purchase expires without being exercised.

If an organization is considering a project that has not been finalized but will follow a tight timetable once approved, the firm may elect to acquire an option on the time-critical elements. For example, if the construction of a new factory is being considered, a suitable piece of land may be optioned. In exchange for a fee, the interested firm might be given an option to purchase the property at an agreed price. If the organization decides not to proceed, it is not obliged to purchase the land, but it will forfeit the option fee.

Buying Capacity

In a fast-moving business environment, it may not be possible to predict the precise numbers of services, products, or components that a supplier will be required to produce. In cases where the volume can be estimated, but the exact requirement mix is unknown, firms may reserve portions of suppliers' capacity. This reduces the uncertainty and risk associated with insufficient capacity and potential lost orders. The supplier is also assured of either production volume or payment for idle facilities. The trade-off is the possibility that the reserved capacity, for which the organization must pay, may not be needed.

Decision Tree Analysis

A typical capital acquisition purchasing decision can be utilized to demonstrate how decision tree analysis technique works. Consider the following scenario:

A purchasing manager is trying to decide whether to buy one machine or two. If only one machine is purchased and demand proves to be excessive, the second machine can be purchased later. Some sales will be lost, however, since the lead time on this type of machine is six months. In addition, the cost per machine will be lower if both are purchased at the same time. The probability of low demand is estimated to be 0.25, the probability of high demand 0.75. The after-tax net present value of the benefits from purchasing the two machines together is $94,000 if demand is low, and $165,000 if demand is high.

If one machine is purchased and demand is low, the net present value is $115,000. If demand proves to be high, the manager has three options. Doing nothing has a net present value of $115,000; subcontracting, $140,000; and buying the second machine, $126,000.[9]

Analysis can be represented in decision tree form, as generically depicted in Figure 8-3. Decisions are represented by a square, and events are depicted by a circle. Branches stemming from a decision node, or square, represent the possible alternatives. Branches stemming from an event node represent the possible scenarios and their associated probability. The far righthand side of the decision tree, the end of each branch, is the expected value or payoff associated with each decision and possible scenario. Figure 8-4 is an illustration of the previously discussed capital acquisition decision.

The decision tree is drawn working from left to right. The solution is derived by working from right to left. From a decision node, select the alternative that has the highest expected payoff. For instance, the second decision node involves a decision between "do nothing," subcontracting, or buying the second machine. Subcontracting has the highest payoff ($140,000), so it is selected, and the others are eliminated. Elimination is indicated by the two slash marks on the branch.

To calculate the event node (EV) expected payoff, multiply the payoff of each branch by the event probability. Add all of the event/probability payoffs together to determine the event node expected payoff. For example, the value of the event node under "purchase two machines" scenario is equal to the probability of low demand (0.25) multiplied by the expected payoff ($94,000), plus the probability of high demand (0.75) multiplied by its expected payoff ($165,000), for an expected value of 147,250.

---

[9]Krajewski and Ritzman, p. 323.

FIGURE 8-3
Decision Tree Framework

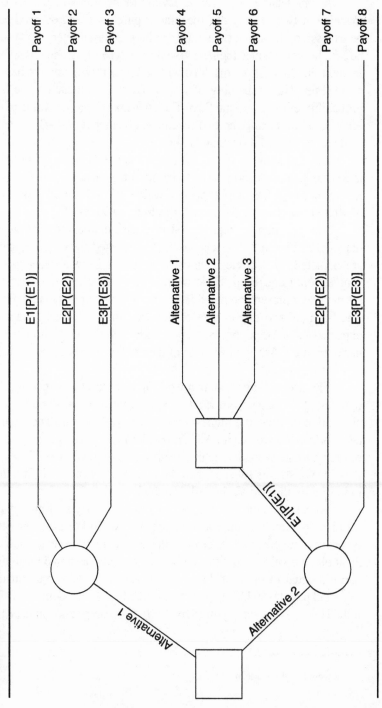

FIGURE 8-4
Capital Acquisition Decision Tree

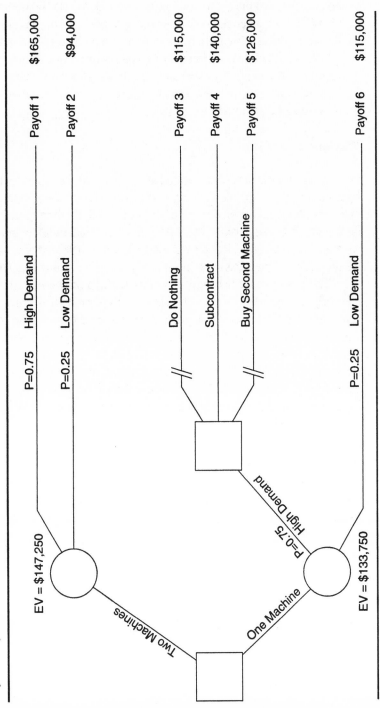

P=0.75	High Demand	Payoff 1	$165,000
P=0.25	Low Demand	Payoff 2	$94,000
	Do Nothing	Payoff 3	$115,000
	Subcontract	Payoff 4	$140,000
	Buy Second Machine	Payoff 5	$126,000
P=0.25	Low Demand	Payoff 6	$115,000

EV = $147,250

EV = $133,750

P=0.75 High Demand

Two Machines

One Machine

When compared to the event node value of $133,750 ([$140,000 X .75] + [$115,000 X .25]) under the "purchase one machine" scenario, it is evident that the initial purchase of two machines will yield the highest expected payoff.

This approach to decision making is a powerful one if the probabilities can be estimated with reasonable accuracy. However, there are relatively few situations in which such estimates are so reliable that the decision maker can entirely trust them.

Purchasing Strategy

Purchasing strategy has been defined as "a set of rules that guides the configuration of the firm's purchasing effort over time in response to changes in competition and the environment so as to permit the firm to take advantage of profitable opportunities."[10] A hierarchical framework of purchasing strategies has been developed by Robert Spekman (1981), that is similar to the hierarchy of business strategies presented earlier in this chapter.

The bottom of the hierarchy is devoted to performance-related strategies that focus on the management of resources. Systems-related strategies, one level higher in Spekman's strategic framework, are directed at the coordination of interdepartmental and inter-organizational functions. Performance evaluation systems, supplier selection, information systems, and value analysis would fall into this category. The highest level involves competition-related strategies. Strategies at this level focus on understanding the dynamics of the buyer-supplier relationship and the conditions present in their respective industries. Supply chain management, international sourcing, and early supplier involvement are examples of competition-related strategies.

Paradigm Shift in the Role of Purchasing

The strategic role of purchasing has evolved significantly since the supply shortages and the influx of foreign competition of the late 1970s. Purchasing has risen from a function viewed as primarily clerical in the eyes of upper management to a vehicle for attaining a competitive advantage in the marketplace.

Similarly, the focus of the purchasing function has changed from pursuing the lowest price to pursuing the highest value. Cost is evaluated on the total cost of ownership, rather than a per-piece price. Relationships between buying and supplying organizations, once described as adversarial, are progressing to

---

[10] Kiser and Rink, 1976, p. 21.

cooperative partnerships. This change in the buyer-supplier relationship stems from recognition of the interdependence and co-destiny that exist between members of the supply chain. The changing dynamics of the inter-organizational relationship have created many new sourcing opportunities that can significantly affect the performance of the firm. The next section will discuss some of the strategic initiatives available to the purchasing function.

Role of Purchasing in Product and Service Design

Concurrent engineering, simultaneous engineering, and integrated product development are just some of the terms used to describe the team approach to the development of new products. Historically, the development of new products followed a linear process through organizational channels, known as the "over the wall" approach. While deemed effective in a relatively static environment, current pressure from consumers and competitors has exposed the major weakness of this approach: time. In response to the need to develop new products quickly, the team approach has been adopted by many progressive firms. The goal is to reduce development time and cost, while improving product quality and performance.

Because 80 percent of product cost and quality is determined by product design, and because an average of 56 percent of each sales dollar is spent on the procurement of production materials, the inclusion of purchasing on the new product development team is imperative for project success.[11] Purchasing can provide information to the team regarding new developments in the supply market, including products and technologies, supplier quality and reliability, and the availability of material.

Purchasing can assume a proactive posture with the supply base, in order to assure the timely delivery of prototypes and component parts. In addition, purchasing can work with the supplier on design issues in order to improve quality and reduce cost. Suppliers are often the best source of information regarding substitute materials or parts that can satisfy the functional requirements of the design at a lower cost.

This proactive approach to new product development provides the purchasing function with visibility regarding future material and service requirements. Purchasing is then capable of negotiating contracts for the suppliers' capacity to ensure the availability of materials and services. Advance design specifications also facilitate the pre-qualification of suppliers in the development of a Qualified Products List (QPL). The list provides

---

[11] Burt, 1989.

timely information on the trade name of the product, model numbers, part numbers, and pertinent information on the manufacturer.

Early Supplier Involvement

The purchasing function is charged with the responsibility of effectively mobilizing and utilizing the external resources available to the firm in the pursuit of a competitive advantage. As such, tapping the expertise available in supplying organizations becomes the responsibility of the purchasing organization. One method utilized to facilitate this process is known as Early Supplier Involvement (ESI).

The most common application of this practice is in the area of new product development. Qualified suppliers are asked to become members of the new product development team. They offer technical and managerial expertise in the areas of product and process technology. Their input can be of significant value with respect to quality, technology, pricing, delivery, material specifications, material tolerances, product standardization, process improvements, product improvements, packaging, inventory, and transportation requirements. By including the supplier in the process, the transition from product design to full-scale production will be much smoother.

While ESI is most often utilized in the area of new product or service development, it should not be limited to this application. Suppliers can provide valuable input in a variety of strategic operating initiatives aimed at process improvements in the supply chain. Electronic Data Interchange (EDI), Just-In-Time deliveries, recycling programs, and cost reduction programs are some examples of areas in which ESI can benefit the buying firm.

Honeywell has been using the team approach and ESI since 1989 in its Air Transport Systems Division. The primary purpose of its commodity teams was to bring suppliers into the new product development process at the blueprint stage. According to Richard M. Grady, C.P.M., technical commodity leader for the division, the team approach has been extremely successful:

> Because of teaming, we don't have as many blueprint change orders coming from engineering and production. Having the suppliers involved up front also allows us to get parts in on time. We've cut lead times in half. [12]

Legal Issues

A common concern regarding the inclusion of the supplier on the new product development team is the issue of confidentiality. Firms are concerned

---

[12] Murphree, 1994, p. 27.

about the possible leakage of proprietary information, which could damage their competitive advantage in the marketplace. The potential and likely consequences of such an occurrence must be evaluated and compared to the potential benefits that can be derived from a close collaboration.

One preventative measure the buying firm can use to minimize the security risk involves requiring suppliers to sign a legal document called a confidentiality agreement. By signing this agreement, the supplier basically agrees not to disclose any information it acquires through its involvement in the new product development effort. If the supplier violates the agreement, it will face some form of penalty.

The legal system provides other vehicles designed to protect the rights of the developer, including patents, copyrights, royalties, trademarks, and licensing. Their applicability depends on the type of product under consideration and the rights of the developer under the law.

A patent gives exclusive privilege to the patentee to make, use, and sell the patented article for 17 years without the threat of competition. Once the patent has expired, the market is open to the entry of competitors. In recent years, numerous patents have expired on prescription drugs, allowing the entry of generic drugs at significantly lower cost to the consumer. Patents are designed to protect the developer for a long enough period of time to allow for the recovery of development costs.

In the sourcing arena, patents may present both opportunities and obstacles for the purchasing organization. Purchasing professionals should pursue patents on all customized products developed by suppliers while under contract with the buying firm. Steps should also be taken to protect the buying organization from patent infringement when utilizing a product that may be patent-protected. Inclusion of a patent indemnity clause in purchasing contracts is a form of protection against possible damages from a liability suit.

While patents apply to products or processes, copyrights are designated for written works. Copyrights protect the rights of the author to publish and sell the original work for the life of the author plus 50 years. Royalties are payments made to the holder of a copyright or patent for the privilege of publishing work or utilizing patent rights. A licensing agreement grants the licensee the right to use the patented invention for a fee and continuing royalties. A license can only be granted by the patent holder.

Trademarks are reserved to protect the usage of a name, symbol, or design for an indefinite time period. The trademark must be registered with the U.S. Patent Office by product category. Trademarks are often utilized to protect a recognized marketing tool. Examples include the "golden arches" of McDonald's and the Coca Cola logo.

Issues in International Purchasing

As previously discussed, the development of international trade agreements and the end of the Cold War have served to develop a new world economy fostering the transfer of goods and services across international borders that were once restricted. In addition, infrastructure support and technology improvements in the telecommunications and transportation industries have reduced the difficulty of conducting business from a distance.

Benefits
This seemingly unrestricted access to new markets has presented the purchasing environment with a host of potential sourcing options, which present both potential benefits and problems.

Potential benefits available from developing international sources of supply include the access to high quality raw materials, component parts, sub-assemblies or original equipment manufacturers (OEM). Lower cost is another potential benefit derived from the access of regional economies to raw materials or reduced labor costs.

Another potential benefit is the establishment of countertrade arrangements that provide access to markets that would otherwise be closed to the firm. Briefly, countertrade agreements involve an exchange of goods and services without a currency transaction. This form of purchase and sale is extremely helpful in opening the channels for marketing and sourcing in countries with a currency that is not easily converted on the world market.

Difficulties
Implementing an international sourcing strategy presents a challenge stemming from the geographic distance and differences in culture, legal systems, and political philosophies. The geographic distance presents logistical coordination issues, which need to be addressed to assure a smooth flow of goods and materials. Coordination is hampered by the different time zones, which affect communication capabilities. With the wide spread adoption of facsimile (fax), however, this issue has been minimized. The distance between the buyer and supplier also complicates the transportation of goods. The longer the supply chain, in terms of distance, the higher the variability in delivery accuracy.

Cultural differences that influence interorganizational relationships include religion, education, language, values, attitudes, traditions, customs, arts, laws, politics, and social organizations. These cultural differences manifest themselves in differing business practices. For instance, in the United States it is customary to expect payment after the delivery of goods or services. In other countries, the supplier expects to be paid in advance. In some countries a verbal agreement or handshake is all that is required to establish a contract, eliminating the need for formal documentation and signatures of commitment. In the purchasing environment, the concept of accepting bribes and the appropriateness of doing so is also culturally determined. It may be considered a standard business practice in some countries.

Language barriers contribute to the difficulty of developing an international sourcing network. Standard communications and negotiations are affected by the differences in both verbal and non-verbal communications. It is not sufficient to be able to verbally communicate with your supplier in his or her native language. The buyer must also understand non-verbal behavior and its meaning in order to establish effective communications.

Another barrier to international sourcing is the lack of widely adopted international standards. The United States Trade Act, passed in 1979, established the international standards code as a part of the Multilateral Trade Agreements. Nations currently committed to utilizing this code include Canada, the European Economic Community, Japan, New Zealand, the Nordic countries, and the United States. Progress is being made in this area through the adoption of standard communication formats for Electronic Data Interchange, and through quality standards established by ISO 9000.

International Politics

Political and regulatory issues are also potential sources of problems in international sourcing. Prior to establishing a sourcing arrangement with a supplier from another country, the buyer must consider the stability of the host government, tariffs, regulations, and the influence of cartels.

An analysis of the political stability of the host country would include a thorough investigation of that region's political history, the form of government, the nature and tenure of current political leadership, the level of civil disorders, the presence of strikes, the presence of conflicts both domestic and abroad, and the stability of the monetary system. The rate and erratic nature of inflation present in a country may lead to political instability, negating any strategic advantage that may have been gained by sourcing from that region.

Another monetary concern in international sourcing is the presence of tariffs imposed on goods exported to or imported from a country. A tariff is a rate of duty placed on the value of the material. It represents an additional cost of doing business in the host country, and it must be considered in the total cost calculations. The presence of cartels can also influence the price and availability of raw materials. Cartels represent the combined interest of a number of producers that act as a single entity in order to gain a monopoly position in the market. In addition to controlling price and availability, cartels control the conditions of sale and allocate market share among the producers. The Organization of Petroleum Exporting Countries (OPEC) is one of the most powerful cartels operating in the world economy. Because of the nearly universal dependence on petroleum and petroleum-based products, OPEC can significantly influence the world economy.

To develop uniformity in regulations that apply to international sales transactions, the Convention on Contracts for the International Sale of Goods has developed a legal standard referred to as CISG. CISG is an attempt to develop an acceptable compromise between the diverse legal systems, philosophies, and business practices in the U.S., Europe, and Latin America. It must be noted that some of the doctrines of CISG are in conflict with those specified by the Uniform Commercial Code. CISG has been widely adopted, and it is expected to have a positive effect on international sourcing.

Financial Considerations

The exchange of goods and services among members of the supply chain is governed by financial transactions. In the international sourcing arena, the transaction process is complicated by different currencies and the documentation required to process the transaction. Of importance to procurement professionals is the total cost of securing the needed goods and services. Part of the cost determination is dependent on the currency exchange rates. The purchasing professional must determine in which currency to negotiate payment—domestic currency, the host country's currency, or a third currency. The decision should be based on the length of the transaction and the volatility of the respective currencies.

Payment between international partners can also be handled by a letter of credit, which is a legal instrument that requires the buyer's bank to pay the supplier upon fulfillment of the contract and presentation of the appropriate documentation. The bank acts as an intermediary in this transaction, billing the buyer for the cost of the purchase and a service fee. This type of transaction benefits both the buyer and the supplier. The buyer obtains the goods without an initial cash outlay. The supplier is assured of payment as long as it fulfills its contractual obligations.

Drafts represent another form of payment used in international transactions. The draft process is initiated by the supplier, which requests payment from the buying organization to the supplier's bank. There are three categories of drafts: an arrival draft, specifying payment upon the arrival of the goods; a sight draft, specifying payment upon presentation of the draft; and a time draft, specifying payment at a predetermined point in time.

A transaction of a different nature involves the transportation of the physical goods from the supplier to the buyer. Air and ship are the primary modes of transportation utilized in international shipments. Air is usually reserved for items of high value and a fragile nature, with weight and space requirements that justify the premium cost of air transport.

Part of the total cost of international sourcing is the cost associated with insuring the cargo during transit. The terminology utilized in these transactions stems from the maritime period, and understanding the terms may require the expertise of an experienced insurance agent or broker. A glossary of relevant international shipping terminology is included at the end of this chapter.

SUMMARY

Effective purchasing planning involves a clear determination of the goals and objectives of the organization and an appropriate timetable for their accomplishment. The process is a top-down procedure, starting with strategic level initiatives and working down to decisions made at the daily operational level. The process must be considered dynamic, with reviews and updates performed on a regular basis. The planning is dependent on the quality of information obtained from the macroeconomic environment, including aggregate forecasts, supply and demand of major commodities, inflation, interest rates, general economic trends, and international trade issues. The purchasing professional acts as the manager of external manufacturing operations, and his or her role in the acquisition planning process is critical to achieving effective internal operations.

KEY POINTS

1. The planning process involves the determination of the organization's mission, goals, and objectives. Planning is the act of determining a proposed set of actions to achieve the organizational goals and objectives and fulfill the mission.

2. There are three levels of planning: strategic, tactical, and operational. They are distinguished by the length of their planning horizon.

3. Effective purchasing strategies are designed to facilitate the accomplishment of the organizational goals, objectives, and mission. The purchasing environment must be linked to the entire organization, and its decisions must be made within this wider context.

4. Decision making consists of three steps: defining the problem, developing and evaluating the alternatives, and implementing the chosen alternative.

5. The purchasing planning process is a dynamic process that should involve periodic review and revisions.

6. Hand to mouth buying, buying to requirements, forward buying, speculation, volume purchasing agreements, life of product, just-in-time, and consignment are among the strategies available to meet the projected supply requirements of the organization.

7. Methods of implementing purchasing strategies include hedging, spot buying, dollar averaging, contracting, escalation and de-escalation clauses, multi-year contracting, line of product contracts, future delivery, options, and buying capacity.

8. Decision tree analysis is a quantitative technique for evaluating alternative solutions with different potential outcomes and probabilities. It can be used in problem solving, or evaluating multiple opportunities.

9. Purchasing strategy can be viewed as a hierarchy of strategies, including performance-related strategies, systems-related strategies, and competition-related strategies.

10. Early Supplier Involvement (ESI) is a strategy designed to include the supplier in the process of developing new products, services, or processes.

11. Access to high-quality sources, lower cost, and new markets through countertrade arrangements are some of the benefits associated with international sourcing.

12. Implementing an international sourcing strategy is hindered by geographic distance, and by differences in culture, legal systems, and political philosophies.

GLOSSARY OF INTERNATIONAL SHIPPING DOCUMENT TERMS

Bill of Lading. This is a receipt for goods provided by the land carriers on both ends of an ocean movement.

Booking Request. This is a form used to reserve shipping space aboard a liner vessel.

Charter Party. This is a lease document between a charterer and a ship owner, used in chartering a tramp vessel.

Delivery Receipt. This serves as proof that the carrier delivered the goods according to the bill of lading.

Dock Receipt. This is an acknowledgment of the goods delivered to the outbound dock.

Import License. This document grants permission to import goods into the U.S.

Ocean Bill of Lading. This is a receipt for goods delivered to the ship for movement.

Shipping Permit. This form serves as proof to the pier personnel that the shipment may be received at the dock.

Through Ocean Bill of Lading. This is used whenever more than one ship line is involved in the movement of goods.

SUGGESTED READINGS

Dill, Thomas F. "Ten Years of Inbound." *Inbound Logistics,* January 1991, pp. 18-21.

Drucker, Peter F. *The Practice of Management.* New York: Harper & Brothers, 1954, pp. 65-83.

Elderkin, Kenton W. and Warren E. Norquist. *Creative Countertrade.,* Cambridge: Ballinger, 1987.

Murphree, Julie. "Top Brass Polished in Purchasing." *NAPM Insights,* February 1991, p. 25.

"Predicting the '90s in Purchasing." *NAPM Insights,* January 1991, pp. 22-23.

"Winning With Upper Management." *NAPM Insights,* October 1993, p. 33.

Wisner, Joel D. "Forecasting Techniques for Today's Purchaser." *NAPM Insights,* September 1991, p. 23.

INDEX

An *n* following a page number (as in 115n) refers to a footnote at the bottom of that page.

A

Accounting/finance, purchasing interface with, 78–79
Accuracy
    of a decision tree payoff, 196
    of forecasts, 31–34
    measuring, 29–30
Acquisitions, 45
Adams, Samuel, 164n
Ad hoc committees, 88
Ad valorem duty rate, defined, 63
Advantages. *See* Benefits
Aetna Life and Casualty, 72
Airbus (France), 55
Alexander, Earl W., 41n
Allen, Robert E., 173
Alliances
    classification of, in partnering, 106
    dissolution of, 115–16
    strategic
        expanded partnering as, 106
        importance of communication in, 96
        with suppliers, 93
    supplier, 104–5
    See also Relationships
American Hospital Supply, 109
American Insurance Association, 119
American National Standards Institute (ANSI), standard for electronic data interchange, 137
American Production and Inventory Control Society (APICS), 121
    information from, 22

American Telephone & Telegraph (AT&T), 53, 68
    environmental strategy at, 158
    toxic waste reduction at, 172–73
Analysis
    break-even, 53
    Bureau of Economic Analysis, 12
    cost-benefit, 183
    decision support systems for, 145
    decision tree, 192–96
    for developing a business strategy, 48
    external, strategic planning process, 55–56
    internal, strategic planning process, 56, 58
    opportunity cost, 183–84
    SWOT, 58
    time series, 26–29
    "total cost-of-ownership", 129
    "what-if"
        with computers, 134
        computers for, 129
        with spreadsheets, 131
    See also Decision making; Value analysis
Andel, Tom, 171n, 172n
Anderson, Eric R., 159n, 161n, 164n
Apple Computer
    restructuring at, 47
    strategic alliance of, 85
Application software, 152
Arco Oil and Gas, 109
Arizona Department of Transportation, 94
    early supplier involvement at, 99
Army and Air Force Exchange Service, 68, 80
Arthur D. Little, poll on corporate citizenship, 158

Artificial intelligence, 6, 146–47
  and decision support systems, 145
Associations
  electronic data interchange, 139
  identifying international suppliers
    through, 62
  professional, 120–23
    information from, 22–24
  representing the organization through,
    119–23
  representing the organization to, 93
  trade, 119–20
Austad Company, forecasting at, 11, 185
Australian Institute of Purchasing and
    Materials, Ltd., 121
Authentication, of computer transmissions,
    149–50
Automation, impact of, on purchasing, 2, 128

B

Baker, R. Jerry, 155, 156n
Balance of payments, 55
Balance of trade, 55
Bales, William A., 180
Banks, identifying international suppliers
    through, 62
Bar coding, 6, 144
  to improve purchasing and logistics
    efficiency, 84
Barter, defined, 61
Baseline
  for economic indexes, 14–15
  for forecasting, 10–11
Basic alliance, classifying partnership in
    terms of, 108
Basic partnering, 105–7
Batch data processing, defined, 153
Bayer, environmental strategy at, 159
Benefits
  of alliance relationships, 110–13
    summary, 112
  classifying partnerships in terms of,
    107–8
  of computerization, 128
  of EDI implementation, 139–40
  of good supplier relationships, 93–95

  of international markets, 200
  in international trade, 54–55
  of a reduced supply base, 116
  value-added, in a business alliance, 109
Bergstrom, Robin P., 158n, 160n, 167n
Berndt, Roger, 89n
Beta, smoothing parameter, 28
Bill of lading, defined, 205
Bishop, Nancy, 145n
Blanket orders, 186
Booking request, defined, 205
Bose Corporation
  alliance maintenance at, 114
  JIT II, 110
Bracker, Jeffrey, 41n
Brainstorming, to generate alternative solu-
    tions to problems, 183
Bramble, Gary M., 146n
Break-even point, 52–53
Bretz, Robert S., 22n
Bribery, culturally acceptable, 201
Bristol-Myers Squibb, "Suppliers Guide to
    Purchasing", 76
Brokers, in international trade, 62
Buckhorn, Inc., 171
Budding, Gonad, 16n
Burt, David N., 197n
Business alliance
  benefits of, 109
  classifying partnerships in terms of, 108
Business Conditions Digest, 20, 21–22
Business strategies, 48–49
  hierarchy of, 42–50
Business structure, evolution of, 84
Buyback, 61
Buying, strategies for, 185–89
  buying capacity, 192
  cooperative, 123
  implementation procedures, 189–96
  inflation versus inventory carrying
    costs, 18

C

Calvin Klein, differentiation strategy at, 48
Capacity, buying to assure supply, 192
Capacity utilization, 19

Capital investment, 100
Carbone, James, 140n
Carlton, Donald D., 129n
Carson, Patrick, 157n, 158n
Cartels, 202
Carter, Joseph R., 90n
Catalogues, electronic, 144–45
Causal forecasting, 29–30
Census, *Bureau of, Statistical Abstract of the United States,* 21
Center for Advanced Purchasing Studies (CAPS), 113
Central processing unit (CPU), defined, 153
Central tendency, measure of, 30
CEOs/Presidents' Perceptions and Expectations of the Purchasing Function, 180
Certification
    of environmental compliance, 169–70
    of suppliers, 103
Certified Professional Purchaser (C.P.P.), 121
Certified Purchasing Manager (CPM), 120
Chambers of commerce, identifying international suppliers through, 62
Champy, James, 80n, 133n
Change
    adaptation to, 50
    calling for dissolution of alliances, 115
    and supplier relationships, 94
Charter party, defined, 205
Charter rate, defined, 64
Chemical Manufacturers Association (CMA), 161–62
    environmental guidelines of, 171
Chevron Petroleum, 159
Chrysler Corporation, 94
    joint venture with Mitsubishi, 45
Clark, Bette, 146n
Clean Air Act, 168
Coca Cola, Inc., value-added networks at, 141
Coincident indicators, 13
Commerce, Department of, U.S., 13
    Bureau of Economic Analysis, 12–13
    identifying international suppliers through, 62

use of Implicit Price Deflator, 17
Committees
    ad hoc, 88
    standing, 89
    as work groups, 87
Commodity management, at Southern Pacific Lines, 89
Communication
    in just-in-time manufacturing, 188
    for sourcing in international markets, 200–201
    with suppliers, 96
    and quality of service, 94
    vehicles for, 68–71
Comparative advantage, 54–55
Compensation, in countertrade, 61
Competition
    changing role of, 124
    perfect, 54
Compound duty rate, defined, 64
Computer Aided Design (CAD), 129
Computers, 6
    for direct ordering by users, 72
    and evolution of purchasing, 2
    impact of, on purchasing, 127–54
    overview, 148–52
Computer system, defined, 153
Concentration, corporate strategy of, 42, 43–45
Concurrent engineering
    role of purchasing in, 197–98
    in a strategic alliance, 110
    See also Early supplier involvement
Confidentiality agreement, 199
    in early supplier involvement, 99
    See also Security
Conflict
    in a functional silo structure, 86
    between internal customers and suppliers, 72–73
    role of purchasing in resolving, 72
Consensus, for problem resolution, 91
Consignment, 189
Consuls, identifying international suppliers through, 62
Consumer Price Index (CPI), 13
    as a measure of inflation, 16

Contingency plans, 182–83
  for meeting change due to natural dis-
    asters, 34
Continuous improvement, 97, 103–4
  with just-in-time manufacturing, 187
  for maintaining an alliance, 114
Contractionary cycles, business planning
    during, 46–47
Contractionary strategies, corporate, 46–47
Contract Management Association, 121
Contracts
  Contracts for the International Sale of
    Goods (CISG), 202
  customs covering, in international mar-
    kets, 201
  evergreen, 187
  financing function involvement, 78–79
  futures, 190
  multi-year, 191
  national, negotiating, 72
  in purchasing planning, 190
  review by the legal department, 81
Cooperative arrangements
  between buyer and supplier, 196–97
  with competitors, 124
Cooperative buying, 123
Copyrights, 199
Corning, Inc., 68, 99
Corporate Environmental Affairs Group, at
    First Brands, 159
Cost
  of electronic data interchange imple-
    mentation, 141
  of noncompliance with hazardous
    waste regulations, 166
  price versus value, 196–97
Cost, Insurance and Freight (CIF), defined, 63
Cost and freight (C&F), transportation
    agreement, 63
Cost-leadership strategy, 48
Cost of living index. *See* Consumer Price
    Index
Cost-push inflation, 18
Cost reduction
  and early supplier involvement,
    100–101, 198
  from EDI implementation, 139–40

from environmental responsibility, 160
from waste reduction, 173
Counterpurchase, 61
Countertrade, 61
  in international markets, 119
    benefits of, 200
Covey, Stephen R., 49n
Creative Countertrade, 61
Creative thinking, in decision making, 183
Credibility
  establishing, 73
  among functions, 72
  in relationships with manufacturing, 82
Cuelzo, Carl M., 10n
Culture
  corporate
    and long-term performance, 49
    risk of conflict in mergers, 46
  functional silo, 86
  and sourcing in international markets,
    200–201
Cumulative sum of forecast error (CFE),
    29
Current requirements, buying to, 185–86
Customers
  internal, 1, 68
  level of demand of, 11–12
  satisfaction of
    role of purchasing, 2, 3
    and supplier performance, 4
Customer service, as an organizational
    goal, 72
Customs broker, defined, 64
Cycles
  contractionary, corporate behavior dur-
    ing, 46–47
  in independent demand, 10
  planning over product lifetime, 185
  purchasing, 135
    reducing clerical effort with com-
      puters, 134
  shortening, by sharing forecasts, 36
Cycle time
  in redesign, 101
  reduction in, with supply chain manage-
    ment, 117
  shortening, with electronic data inter-

change, 143, 144
Cyclical indicators, 21–22

D

Data
adjustment of, using indexes, 13–14
from commercial sources, 24
computers for access to, 128
computers for processing and storing, 129
defined, 153
for economic indicators, 13
external, 20
government sources of, 20–22
international, 22
measures of dispersion of, 30
security of, in value-added networks, 141
sources of, for forecasting, 19–24
Database management systems, 131–32
defined, 153
in the purchasing process, 134
Davidow, William H., 85n
Davis, Tom, 117n
dBase IV, 152
Decision making, in planning, 183–84
Decision support systems, 6, 145
analyses for buyers, 134
computers for, 129
Decision tree analysis, 192–96
Deep-well injection, for disposal of solid waste, 162
Deficit, government, 60
Delivery receipt, defined, 205
Dell Computer Corporation, 32
Delphi method, 31
Demand
dependent and independent, 10
effect of marketing strategies on, 79
Demand management, 11–12
Demand-pull inflation, 18, 60
Dependence
mutual, with suppliers, 94, 106
on a reduced supplier base, 98
on suppliers, in single or limited bases, 116
Deregulation

and joint ventures, 45
of transportation, 59
Design
and cost of manufacturing, 182
reducing cost at the level of, 197
supplier involvement in, 100
redesign, 101
Desktop publishing, 130
Development
of alliance relationships, 113–16
of suppliers, 102
See also Product development
Diagonal communication, 69, 71
Differentiation strategy, 48–49
Diffusion index, 22–24
Digital Equipment Corporation, electronic catalogue, 144–45
Diminishing Marginal Returns, Law of, 52
Diminishing Marginal Utility, Law of, 51–52
Discount rate, 18–19
Distributed data processing, defined, 153
Diversification
corporate strategy of, 42, 43
related, examples of, 46
unrelated, examples of, 46
Divestiture, to improve corporate performance, 47
Divisions, functional co-location in, 87
Dock receipt, defined, 205
Dodge Corporation, F.W., 13
Dollar averaging, in purchasing planning, 190
Dow Chemical, environmental strategy at, 158
Dow Jones, 13
Downsizing, 46
Drafts, for international payments, 203
Drucker, Peter F., 177, 177n
Dumping, in international markets, 55
Dunn & Bradstreet, 13
Duty rate, kinds of, defined, 63

E

Earl Scheib (firm), cost-leadership strategy at, 48

Early purchasing involvement (EPI), in product development, with need for capital investment, 100

Early supplier involvement (ESI), 99–101, 196, 198
    benefits of, 36
    in change, 94
    computer modeling of new products for, 129

Economic Analysis, Bureau of, *Survey of Current Business,* 20–21

Economic conditions, effect of, on forecast accuracy, 33–34

Economic indicators
    defined, 12–13
    forecasting from, 12
    summary of sources, 14

Economic order quantity (EOQ), computerized model for determining, 133

Economic Report of the President, 20, 21

Economics, general issues, in strategic planning, 58–60

EDS, 45

Educational and Institutional Cooperative Service, Inc., 123

Effectiveness
    and forecasting accuracy, 34–36
    improvement in ordering processes, 97

Efficiency
    computerization for, 128
    from good supplier relationships, 94–95
    from specialization, 84

Elasticity, price, 50–51

Elderkin, Kenton W., 61, 61n

Electronic Data Interchange (EDI), 6, 136–44, 198
    and evolution of purchasing, 2
    implementation steps, 142–43
    to improve purchasing and logistics efficiency, 84
    in just-in-time manufacturing, 188
    for purchase order handling, 79–80
    standards for, 201

Electronic Funds Transfer (EFT), for paying for catalog orders, 145

Electronic mail (E-mail), 130

Eli Lilly, restructuring at, 47

Ellram, Lisa M., 113n, 118n

Embassies, identifying international suppliers through, 62

Employment Act of 1946, 60

Empowered teams versus consensus management, 91

Encryption, of computer transmissions, 150

Encyclopedia of Associations, 123

Energy recovery, in treatment of solid waste, 162

Engineering/design/R&D, interaction with purchasing, 75–76

Environment
    concerns about, 6
    environmental protection/prevention (EPP), 159
    external, relationship of the business entity to, 49–50
    nature of, in forecasting, 25
    operating
        and level of demand of, 11–12
        and supplier relationships, 94

Environmental certification, 168–69

Environmental Management Systems (EMS), standards for, 170

Environmental Protection Agency (EPA), 163
    enforcement by, of federal regulations, 165–66

Ericsson, Holger, 145n

Error
    computerization for reducing, 128
    from failure to enter data in computer systems, 133
    in forecasts, measuring, 29
    in receiving, bar coding to reduce, 144

Escalation clauses, in purchase contracts, 191

Ethyl Corporation, 186
    just-in-time manufacturing at, 188

European Community (EC), green certification in, 169

European Economic Community (EEC), 54

Evaluation, of forecasts, 35–36

Evans-Correia, Kate, 72n
Evergreen contracts, 187
Evolution
    of business structures, 84–91
    of the purchasing function, 2
Exchange rate
    and cost of goods, 202
    defined, 55
    effect of, on forecast accuracy, 34
Expanded partnering, 106–7
Expansionary strategies, corporate, 42–46
Expectations, of an alliance, clarifying in
    writing, 114
Expected value, decision tree analysis,
    193–94, 196
Expert assessment, for forecasting, 31
Expert systems
    artificial intelligence use, 146
    potential applications in purchasing,
    147
Exponential smoothing, 28–29
Export trade companies, identifying inter-
    national suppliers through, 62
External resources, role of purchasing in
    managing, 42
Exxon, oil spill, *Valdez,* 160

F

Fair trade, 61
Fearon, Harold E., 121n, 180
Federal Register, 166
Federal Reserve Board, 13
    Federal Reserve Bulletin, 21
    Industrial Production Index of, 17
    management of the economy by, 18–19
Federal Reserve Bulletin, 20, 21
Federated Department Stores, 46
Feedback
    reviewing purchasing strategies, 184
    soliciting from suppliers, 97–98
Files, computer, 150–51
    defined, 153
First Brands Corporation, environmental
    strategy at, 158
Fisher, Marshall L., 11n, 32n, 35n

Fit
    environmental, in strategic planning, 41
    strategic, 49–50, 58
"Five Force Model", 56
Fixed costs, defined, 52
Florimo, Robert, 146n
Focus
    on balance of power, "Five Force
        Model", 56
    strategic, in purchasing, 49
Focus strategy, 48–49
Ford Motor Company
    computerized receiving at, 79–80
    joint venture with Mazda, 45
    receiving system, 133
Forecasting, 7–38
    annual cycle in, 185
    computers for, 129
    defined, 7
    internal, 24
    linking purchasing strategies with,
        184–85
    of profits, 182
Forward buying, 18, 186
Fragmented industries, 54
Framework, for strategic planning, 50–55
Free trade, defined, 61
Full employment, government policy and,
    60
Functional co-location, in divisions, 87
Functional silos, 86
Functional specialization, 84–85
Functional strategy, 49–50
Future delivery contract, 192
Futures contract, 190

G

GE Medical Systems, integrated supplier
    program, 117–18
General Agreement on Tariffs and Trade
    (GATT), 54
General Electric Company
    of Australia, supplier feedback at, 97
    development of value analysis at,
        101–2

supplier continuous improvement, 103
General Mills, development by, of Olive
    Garden, 46
General Motors, 45
    divestiture by, 47
G & F Industries, 110
Global sourcing, and accuracy of forecasts,
    34
Glossary
    basic computer terminology, 152–54
    international procurement terms, 63–64
    international shipping document terms,
        205
Goals
    business, customer service, 3
    environmental, standards in the
        European Community, 169
    organizational, 58
    and user satisfaction, 81
    strategic, 177
Gold standard, defined, 55
Government, budgeting by, 60
Grady, Richard M., 198
Graphics, computers for, 129
Green Cross, certification system for envi-
    ronmental compliance, 169–70
Green movement, 157–60
    certifying environmentally friendly
        practices, 168–69
    effect of, on purchasing, 171–73
Green Report Card, 170
Gross Domestic Product (GDP) deflator, 17
Gross National Product (GNP) deflator, 17
Group purchasing associations, 123
Guess, 45
    differentiation strategy at, 48
Guidant, spin-off from Eli Lilly, 47

H

Hale, Roger L., 129n
Hammer, Michael, 80n, 133n
Hand to mouth buying, 185
Hardware, computer, 148
    defined, 153
Harrigan, Kathryn Rudie, 45n

Hatchett, Ed., 144n
Haver, Maurine, 15
Haver Analytics, 15
Hazardous Communication Standard
    (OSHA), 165
Hazardous material
    defined, 160–61
    methods for disposal of solid waste,
        162–64
    in waste, responsible disposal of, 155–74
Hazardous Materials Transportation Act,
    165
Hazardous Material Technology, Office of,
    167
Health Resource Institute of Los Angeles,
    123
Heberling, Michael E., 75n
Hedging, as a commodity purchase strategy,
    189–90
Hendrick, T.E., 113n
Herman Miller, Inc.
    environmental strategy at, 158
    waste disposal at, 164
Hewlett-Packard (HP)
    early supplier involvement at, 99
    process mapping at, 117
Hierarchy
    of business strategies, 42–50
    in expanded partnering, 106
    in planning, 178, 180
    or purchasing strategies, 196
Hightower, Shakira, 47n
Hollow corporation, 85
Honda
    supplier development at, 102
    supplier training at, 99
Honeywell
    Air Transport Systems
        early supplier involvement at, 198
        supplier rating, using database man-
            agement systems, 132
    Space and Aviation Systems, 178
    supplier certification, 103
    supplier performance expectations, 104
Horizon, planning/time, 177–203
    for forecasting, 25

and product success, 5
summary, 27
Horizontal communication, 70
between functions, 69
Hyperinflation, 60
Hyundai Motors, cost-leadership strategy
at, 48

I

IBM
restructuring at, 47
strategic alliance of, 85
Image, purchasing's, benefit of computeri-
zation to, 128
Implementation plan, factors in, 58
Implicit Price Deflator, 17
Import license, defined, 205
Incineration, for treatment of solid waste, 162
Incremental paper trail, in electronic data
interchange, 139
Indexes
composite, leading economic indicators, 12–13
concept of, 13–15
Consumer Price Index, 15–16
diffusion index, 22–23
Industrial Production Index (IPI), 17–19
price, 15–17
Purchasing Managers Index, 23
Industrial Darwinism, 50
Industrial Production Index (IPI), 17–19
capacity utilization factor, 19
Inflation
Consumer Price Index as a measure of,
16
defined, 17–18
effect of, on timing of purchases, 18
international, and sourcing arrangements, 201
and strategic plans, 60
Information
defined for computer use, 153
effect of computerization on, 147
to employees, about hazardous materials, 165

external, purchasing's access to, 178
external contacts for gathering, 3–4
flow of, 68–69
from forecasting, 9–12
on interest rates, sources of, 19
internal, purchasing's access to, 178
provided by the purchasing function, 67
shared, on a computer network, 152
See also Data
Information flows
to functional areas, from purchasing, 74
management of, in the supply chain, 117
Information systems, 196
interactions with, 79–80
See also Management information systems (MIS)
Input devices, defined, 153
Institute of Purchasing and Supply (Great
Britain), 121
Insurance, of goods in transit, international
sourcing, 203
Integrated word/data processing, defined,
153
Integration
in manufacturing, to minimize waste,
168
in product development, 197–98
of purchasing, strategic focus on, 49
Intel Corporation
linkage of purchasing and logistics at, 8.
open door policy for suppliers, 97
Interaction
interdepartmental styles of, 84–91
organizational, types of, 70
Interest rates
discount rate, 18–19
effect of, on forecast accuracy, 33
prime rate, 19
and strategic plans, 59
International Federation of Purchasing and
Materials Management, countries
represented in, 121
International practices
in purchasing, 200–202
in reciprocal supplier relationships, 119
International sourcing, 196

"International Trade Statistics", 22
Inventory
    control of, use of computers in, 132–33
    cooperation of stores to manage, 81–82
    hedging to protect the value of, 190
    management of
        point of sale (POS) system for, 80
        in the supply chain, 117
    minimizing stock levels, 7–8
    reduction of, bar coding and EDI application, 144
"Invisible hand", 54

J

Japan Materials Management Association, 121
JIT II, at Bose Corporation, 110
    steps in implementation of, 111
John Deere Company, 15–16
Joint ventures, 45
Jones, Del, 47n
Joy, Pattie, 47n
Just-in-time procedures, 32, 187–88, 198

K

Katzenbach, Jon R., 89n
Keen, Howard, Jr., 13n, 17n
Keiretsu, 119
    American adaptation, 105, 106
Kirschner, Elisabeth, 160n
Kiser, G.E., 196n
Klein, Philip A., 23n
Kowal, Ronald E., 129n
Kraft Foods, 45
Krajewski, Lee, 46n, 193n

L

Labor Statistics, Bureau of, 13
    cost of price data collection by, 16
    data for the Producer Price Index from, 15
Labor supply
    effect of, on forecasting, 32–33
    full employment and strategic planning, 60
    redistribution of, in contractionary business cycles, 46
Lagging indicators, 13
Landfill, for disposal of solid waste, 162
Leading indicators, 12–13
    diffusion index, 23–24
Lead times, and accuracy of forecasting, 32
Leenders, Michiel, 121n
Legal considerations
    in early supplier involvement, 198–200
    in electronic data interchange, 143
    in hazardous waste disposal, 165–66
    in preference to suppliers, 118–19
Legal function, interaction with, 81
Letter of credit, 202–3
Liability, in hazardous materials/waste management, 166–67
Licensing, of patented products or processes, 199
Linear regression, 29
Linkages
    in division structure, 85
    through electronic data interchange, 143
    of the EPA with RCRA, 166
    among levels of planning, 180–81
    between strategic plan and operating environment, 7–8
    in the value-added chain, 50
L.L. Bean, 36
Local area networks (LANs), 149, 152
    defined, 153
Logistics, 82–84
    of sourcing in international markets, 200–201
Logistics Management, Council of, 121
Lotus 1-2-3, 152

M

M-1, M-2 and M-3, defined, 59
McAllister, Bill, 47n
Macy's, 46
Mailbox, electronic, in value-added networks, 138

Mainframes, overview, 148–49
Maintenance, of an alliance, 114–15
Malone, Michael S., 85n
Management, top, relationship of purchasing with, 75
Management information systems (MIS), 79–80
  defined, 154
Marcus, Jeff, 170n, 171n
Marginal utility, defined, 52
Market demand curve, 51
Marketing, reverse, 102–3, 105
Marketing function, communication with, 79
Market research, 30–31
Markets
  forecasting, 10
  international, 54–55
  nature of, 50–54
  time to, reducing with supplier involvement, 101
Material requirements planning (MRP), computers for, 133
Materials, commodity plans, 182
Material Safety Data Sheets (MSDS), 165
Materials management, 83
  recovery of materials from solid waste, 162
Matrix structure, 85, 87
Mazda, joint venture with Ford Motor Company, 45
MCI Communications, 94
  early supplier involvement at, 99
  electronic mail at, 130
Mead Corporation
  artificial intelligence use, 146
  environmental strategy at, 159
Mean, statistical, 30
Median, 30
Mercedes Benz, differentiation strategy of, 48
Mergers, 45–46
Methods, for forecasting, 26–29
Microcomputers, 149
Minicomputers, 149
Ministry of International Trade (MITI), Japan, 45

Minitab, 30
Mintzberg, Henry, 42n
Missions, government, identifying international suppliers through, 62
Mission statement, 56, 58
Mitsubishi, joint venture with Chrysler Corporation, 45
Mobil Corporation, 47
Mode, of a data set, 30
Modeling, computers for, 129
Monetary policy, and employment levels, 60
Monet Jewelers, 45
Money supply, defined, 59
Monopoly, defined, 53
Moore, D. Larry, 178, 180
Moore, Geoffrey H., 23n
Moore, Steven, 186
Motorola, 91, 94
  computers for "total cost-of-ownership" modeling, 129
  strategic alliance of, 85
  supplier training at, 99
  work team approach at, 90
Moulden, Julia, 157n, 158n
Moving average, 26–27
  weighted, 27–28
MS-DOS, 154
Mules, Glen R.J., 150n
Multilateral Trade Agreements, 201
Multiple regression, 29–30
Murphree, Julie, 178n, 180n, 198n

N

Nabisco, 45
NAPM Insights, 182
National Association of Educational Purchasers, 123
National Association of Hospital Purchasing Management, 121
National Association of Purchasing Management (NAPM), 120
  on corporate citizenship, 155–56
  definition of computer network, 151–52

Dictionary
  definition of supplier development, 102
  definition of supplier partnership, 104–5
  information from, 22–23
  support for EDI standards, 137
National Association of Realtors, 120
National Car Rental, 47
National Electric Manufacturers Association, 120
National Enforcement Investigative Center (EPA), 165–66
National Institute of Government Purchasing, 121
National Trade and Professional Association of the United States, 123
Natural disasters, effect of, on forecasts, 34
Natural language capabilities, 146
Nautica Apparel Inc., 45
Negotiation
  clauses covering inflation in long-term contracts, 18
  Producer Price Index use in, 15–16
Networks
  computer, 151–52
  corporation, of externally purchased activities, 85
Neural networks, 146–47
Newman, Richard G., 102n
Niche, in a focus strategy, 49
Nikon, differentiation strategy at, 48
Node, in process mapping, 117
Nomad (application software), 152
Norquist, Warren E., 61, 61n
North American Free Trade Agreement (NAFTA), 54
Northern Telcom, waste reduction at, 173
"Not in my back yard", 157, 164

O

Objectives, organizational, 58
  supply contribution to, 181
  support of, 180
Occidental Petroleum Corporation, 157

Love Canal contamination, 160
Occupational Safety and Health Act (OSHA), 165
Ocean bill of lading, defined, 205
Offset arrangement, in buyback, 61
Ohmae, Kenichi, 114n
Oklahoma, University of, 180
Oligopoly, defined, 54
Olive Garden, 46
On-site waste disposal, 164
Operational alliance
  classifying partnership in terms of, 108
  emphasis on buyer benefit, 108–9
Operational perspective, 1–2
  in planning, 178
Operations/manufacturing, working relationship with, 82
Opportunity cost, analysis of, 183–84
Options, for purchasing, 192
Oregon, state of, request for proposal (RFP) system, E-mail, 130
Organization of Petroleum Exporting Countries (OPEC), 202
OS-2, 154
Output devices, defined, 154
Outsourcing, 85
  and evolution of purchasing, 2
  of legal department functions, 81
Over the wall approach, product development, 197

P

Pacific Bell, 94
  joint purchaser-supplier training program of, 98
Packaging, reducing solid waste through design of, 171
Paradigm shift, in the role of purchasing, 196–97
Parallel data processing, defined, 154
Partnerships with suppliers, 104–5
  types of, 105–10
Patent, protection of, 199
Patent Office, U.S., trademark registration with, 199–200

Payment, 95
international, 202–3
customs concerning, 201
policy on, 78
Perceived value, to the customer, in a business alliance, 109
Performance requirements, 76
compliance with
database management system for measuring, 132
by suppliers, 93–94
and dissolution of alliances, 115–16
helping suppliers meet, 98–104
responses to violation of, 115
Personal computers (PCs), 149
Philip Morris, 45
Phillips, Stephen, 48n
Planning
administration of, by accounting/finance, 78–79
advanced acquisition, 184
defined, 177
versus forecasting, 9–12
Point of sale (POS) system, 80
Politics
and forecast accuracy, 34
international, and sourcing arrangements, 201–2
Pollution Prevention Act of 1990 (PPA), 167, 168
Porsche, differentiation strategy at, 48
Porter, Ann Millen, 110
Porter, Michael, 48, 55, 55n
Post Office, United States, 46–47
Power, determinants of, 57
"Preventing Pollution Pays", 3M program, 158
Price change clause, 191
Price indexes, 15–17
Prices
changes in, and demand management, 12
equilibrium, 51
relationship of cost to value, 196–97
of supplies, forecasting, 10
Prime rate, 19

Proactive role
in adapting computer technologies to purchasing, 148
of purchasing, 72, 124
in product development, 197–98
in waste management, 168–71
Problem definition, in planning, 183
Process mapping, in supply chain management, 117
Proctor and Gamble, restructuring at, 47
Procurement strategy
international, 61–62
to limit risks in global sourcing, 34
Producer Price Index (periodical), 15
Producer Price Index (PPI), 14, 15–16
Product-delivery process, 7
Product development
cost and profitability estimates, 79
information about materials, 76
integrated, 197–98
interaction with purchasing in, 75–76
involvement of suppliers in, 94
and patent rights, 199
and strategic success of organizations, 4–5
team for early supplier involvement in, 99
waste source reduction considerations, 168
Product life cycle
and demand forecasting, 11
disposal as part of, 167
supplier contracts covering, 187, 191–92
Products
hazardous, responsibility of the producer for, 166–67
unitary elastic, 51
Professional associations, 120–23
list, 122
Profit
in the "Five Force Model", 56
as an objective, 182
Program, computer, defined, 154
Proprietary systems, for electronic data interchange, 138

Public relations/public affairs function, interactions with purchasing, 81
Purchase Connections, 123
Purchasing function
evolution of, 2
identifying and certifying waste disposal sites, 164
Purchasing Management Association of Canada (PMAC), 120–21
Purchasing Managers Index (PMI), 23
Purchasing planning, 177–205
Purchasing strategy, 196–200

Q

Qualified Products List (QPL), 197–98
Qualitative methods, for forecasting, 31
Quality assurance
interface with, 76, 78
international standards for, 201
in just-in-time manufacturing, 187
with a reduced supply base, 116
Quality circles, 91

R

Rademaker, Ken, 160n
Random error, in determining demand, 10
Range, of a data set, 30
Rashid, Sharon, 145n
Rationalization, of the supply base, 116
Real time data processing, 154
Receiving, computer tracking of, 133–34
bar coding, 144
Reciprocity, in supplier relationships, 118–19
Recovery, of hazardous material, 163–64
Recycling, early supplier involvement in, 198
Redesign, supplier involvement in, 101
Red Roof Inns, cost-leadership strategy at, 48
Regulation, of utilities, 53
Rehee, R. Anthony, 102n
Relationships
alliance

development of, 113–16
risks and benefits of, 110–13
collaborative, in multi-year contracting, 191
effect on, of value-added networks, 141
external
emphasis in basic alliance, 108
purchasing's role in, 93–125
role of purchasing in, 123–24
among functional areas, 73–84, 87
interfunctional, 67–84
internal, managing, 67–92
supplier
impact of reciprocity on, 119
means of promoting, 95–98
traditional, with suppliers, 1
See also Alliances
Report on Business (ROB), 22–23
use in commodity planning, 182
Request for proposal (RFP), using E-mail for, 130
Research consortiums, 45
Resource Conservation and Recovery Act (RCRA), 160–61, 166, 168
Resource management, 196
Responsibility
for hazardous materials, 166–67
social, and hazardous waste disposal, 155–74
of a task team, 90
Restructuring, corporate, 47
Reverse marketing, 105
Rework costs, computer monitoring of, 134
Rhodes, Stanley, 170
Rightsizing, 46
Rink, David, 196n
Risks
of alliance relationships, 110–13
summary, 113
in computer use, security, 149–50
of EDI implementation, 140–41
of price or cost changes, 191
reducing with supply chain management, 117–18
Ritz Carlton, differentiation strategy of, 48
Ritzman, Larry, 46n, 193n

RJ Reynolds, 45
Rolex, differentiation strategy at, 48
Rolls Royce, differentiation strategy at, 48
Rothfuss, Robert, 171–72
Roush, Chris, 45n
Rowson, David, 169n
Royalties, 199
Runyon, Marvin, 46–47

S

Sales function, communication with, 79
Sansolo, Michael, 170n
Schnert, Tim K., 129n
Scientific Certification Systems (SCS), 170
Seasonal fluctuations, in independent
    demand, 10
Secondary storage, computers, defined, 154
Security
    in computer use, 149–50
    in early supplier involvement, 198–99
    reisk to, in alliance relationships, 111
Selection, of suppliers, for an alliance rela-
    tionship, 113
Self-directed teams, 90–91
SEMATECH approach, to partnering rela-
    tionships, 105–8
Sensitivity, in interfunctional interactions,
    69
Service contracts, 191–92
Service function, purchasing as, 1
Sheridan, John H., 167n, 173n
Shipping permit, defined, 205
Single sources, 116
    future delivery contracts with, 192
Skupsky, Donald S., 143n
Sloane, David P., 161n
Smith, Adam, 54
Smith, Douglas K., 89n
Social responsibility, corporate, 155–60
Software, computer, 148
    application, 152
    defined, 154
Solid waste, methods of treating, 162
Source reduction, for waste management,
    168

Sourcing, international, and accuracy of
    forecasts, 34
Southern Pacific Lines, commodity man-
    agement at, 89
Special projects, reporting relationships
    for, 85
Specific duty rate, 64
Speculative buying, 186
Spekman, Robert E., 49n, 196
Sport Obermeyer, forcasting with "accu-
    rate response" method, 11
Spot buying, 189
Spreadsheets, 131
Stakeholders, 156
Standard deviation, 30
Standards
    for electronic data interchange, 136–37
    for Environmental Management
        Systems, 170
    international, 201
Standing committees, 89
State, Department of, U.S., risk assess-
    ments by, 34
Statistical Abstract of the United States,
    20, 21
Statistical Analysis System (SAS), 30
Statistical Package for the Social Sciences
    (SPSS), 30
"Statistics of World Trade in Steel", 22
Steelcase Company, waste minimization
    at, 172
Stores/facilities function, communication
    with, 81–82
Strategic alliance
    classifying partnership in terms of, 108
    commitment of supplier in, 110
    See also Alliance
Strategic fit, 49–50
Strategic function, of purchasing, 67
Strategic goals, firm's, and evolution of
    purchasing, 2
Strategic planning
    by the EPA, RCRA Implementation
        Plan, 166
    framework for, 50–55
    organizational, role of purchasing in, 3–5

process of, 41–64
role of forecasting in, 9
Strategies
  business, 48–49
  cooperative, 45
  corporate, 42–47
    support from electronic data inter-
      change, 143
  defined, 41–42
  green, development of, 158–60
  nested, 49
  proactive, for waste management,
    168–71
  purchasing, 196–200
  sourcing in international markets, prob-
    lems with, 200–201
Strengths, weaknesses, opportunities, and
  threats (SWOT), 58
Strikes, effect of, on forecasts, 33
Structural unemployment, defined, 60
Subsidies, competitive advantage created
  by, 55
Sunkel, Jill, 130n
Suppliers
  alliances with, 104–5
  benefits of good relationships with,
    93–95
  buying capacity of, 192
  capability to meet specifications, 78
  commitment required for partnership,
    107
  demands on, with just-in-time manu-
    facturing, 187–88
  development of, 102–3
  fairness to, 73
  international, resources for identifying, 62
  involvement with purchasing, forms of,
    99–103
  quality of, and customer satisfaction, 3
  selection of, 196
  size of base, and quality of relation-
    ships with, 96–97
  supplier policy statement, example, 77
Supply, strategic contribution of, 179
Supply and demand, and price, 50–51
Supply base reduction, 116, 124

and multi-year contracting, 191
and quality of relationships, 96–97
Supply chain
  buyer-supplier relationships in, 197
  management of, 117–18, 196
  sharing forecasts in, 35–36
  and sourcing in international markets,
    200–201
  waste reduction in, with just-in-time
    manufacturing, 187
Supply curve, 51
Support, from the purchasing function, 67
Survey of Current Business, 20–21
Swedish National Association of
  Purchasing and Logistics, 121
Switch transaction, in countertrade, 61
Synergy, of development staffs, supplier
  and purchaser, 101
Systems, for performance evaluation, 196
Systems approach, to supply chain man-
  agement, 117
Systems software, defined, 154

T

Tactical plan, 178
Tactics, for implementing strategies, 58
Tariffs, and costing, 202
Task teams, 89–90
Teams, 87–88, 89–91
  for product development, 197–98
  use in product development at
    Honeywell, 198
Techniques, for forecasting, 24–36
Technology
  development of, by supplier, 101
  effect of, on forecast accuracy, 33
  reducing risk of adopting, 111
Tennant Corporation, 98
  new product modeling at, 129
  suppliers
    certification of, 103
    feedback from, 97
    training program, 99
Texas Instruments
  bar coding application, 144

environmental strategy at, 159
"total cost-of-ownership" modeling at, 129
3M, environmental strategy at, 158
Through ocean bill of lading, defined, 205
Thums Long Beach Company, 109
Timberland Company, 35
Time, for development of alliance relationships, 114
Time average of demand, 10–11
Time-based competitive strategies, 36
  reducing time to market with supplier involvement, 101
Time series analysis, 26–29
Timex, concentration strategy of, 45
Torda, Theodore S., 23n
Total cost, 52
Total revenue, 52
Toxic Release Inventory, 167
Trade Act, United States, 201
Trade associations, representing the organization through, 119–20
Trade barriers, 55
Trademarks, 199–200
Trade secrets, risk to, in alliance relationships, 111
Trading partner agreements, 143
Training
  in computer use, 132
  in hazardous materials management, 165
  purchaser-supplier programs, 98–99
Transportation
  cost of, in managing reusable containers, 171
  of hazardous waste, 164
    liability implications, 167
  interface between purchasing and logistics, 83
  in international markets, 200–201, 203
  trends in, and strategic plans, 59
Transportation, Department of, U.S., enforcement by, of the Hazardous Materials Transportation Act, 165
Trends
  Business Conditions Digest report of, 22
  in independent demand, 10

Trust
  and communication, in supplier relationships, 96
  in developing an alliance, 114
  establishing, 73
  among functions, 72
  in just-in-time manufacturing, 188
  in relationships with manufacturing, 82
  in supplier relationships, 36, 73, 94

U

Uniform Commercial Code
  conflict with Contracts for the International Sale of Goods, 202
  on written contracts, 143
Uniform Hazardous Waste Manifest, 166
Union Carbide, Bhopal industrial accident, 160
United Nations, information business data provided by, 22
United Parcel Service (UPS), 33
User, inputs from, 80–81

V

Vajda, Gary, 168n
Value-added benefit
  in a business alliance, 109
  in a strategic alliance, 110
Value-added chain
  breadth of, and vertical integration, 43
  linkages in, 50
  purchasing in, 72
  role of forecasting in, 9
Value-added issues, freeing time to work with, 128
Value-added networks (VANs), for electronic data interchange, 138
Value analysis, 101–2, 196
  involvement of supplier in, 103
  supplier involvement in, 101–2
  See also Analysis
Variable costs, defined, 52
Variance, of data, 30
Vertical communication, with functions, 69

Vertical integration, corporate strategy, 42
Veryfine Products, Inc., 164
Virtual corporation, 85
VM, 154
Voice recognition, artificial intelligence
    systems, 146
Volume purchase agreements, 186

W

Wall Street Journal, information on inter-
    est rates, 19
Wasik, John, 158n
Waste exchange firms, 164
Waste management
    with just-in-time manufacturing, 187
    magnitude of problem, 161
Waste Management Corporation, 163
Wastewater, hazardous, methods of treat-
    ing, 162
Waters, James A., 42n
Wealth of Nations, The, 54

White Castle, cost-leadership strategy at, 48
Whitman, Earl, 180
Wholesale Price Index (WPI). *See*
    Producer Price Index
Wide area networks (WANs), 152
Wilsher, Peter, 158n, 159
Wisner, Joel D., 11n, 185n, 186n, 188n
WordPerfect, 152
Word processing, 130
Work groups, 87–88
Work teams, 90–91
"World Economic Survey", 22
Wright, John W., 12n, 13n, 15n, 16n, 17n,
    18n

Y

Yacura, Joseph A., 147n

Z

Zaibatsu, 119

Thank you for your order.

If you would like more information about our programs and services, please complete this card and drop it in the mail.

Name _____

Address _____

City _____ State _____ Zip _____

Yes, please send me information about:

☐ Membership
☐ Certification
☐ Seminars
☐ Educational Products

National Association of Purchasing Management ®

Thank you for your order.

If you would like more information about our programs and services, please complete this card and drop it in the mail.

Name _____

Address _____

City _____ State _____ Zip _____

Yes, please send me information about:

☐ Membership
☐ Certification
☐ Seminars
☐ Educational Products

National
Association of
Purchasing
Management

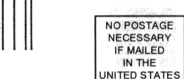

NO POSTAGE
NECESSARY
IF MAILED
IN THE
UNITED STATES

BUSINESS REPLY MAIL
FIRST-CLASS MAIL   PERMIT NO. 105   TEMPE AZ

POSTAGE WILL BE PAID BY ADDRESSEE

National Association of
  Purchasing Management, Inc.
P.O. Box 22160
Tempe, Arizona 85285-9781

National
Association of
Purchasing
Management

NO POSTAGE
NECESSARY
IF MAILED
IN THE
UNITED STATES

BUSINESS REPLY MAIL
FIRST-CLASS MAIL   PERMIT NO. 105   TEMPE AZ

POSTAGE WILL BE PAID BY ADDRESSEE

National Association of
  Purchasing Management, Inc.
P.O. Box 22160
Tempe, Arizona 85285-9781